# Non-linear transport properties of hybrid nanoelectronic devices

Henning Soller

Bibliografische Information der Deutschen Nationalbibliothek

Die Deutsche Nationalbibliothek verzeichnet diese Publikation in der Deutschen Nationalbibliografie; detaillierte bibliografische Daten sind im Internet über http://dnb.d-nb.de abrufbar.



ISBN 978-3-8325-3379-3

Logos Verlag Berlin GmbH
Comeniushof, Gubener Str. 47,
10243 Berlin
Tel.: +49 (0)30 42 85 10 90
Fax: +49 (0)30 42 85 10 92
INTERNET: http://www.logos-verlag.de

# Contents

# Chapter 1

## Introduction

> Simplicity is the sign of truth. That is a guiding principle for my research in theoretical physics.
>
> *(Carlo Beenakker, December 2009)*

Thanks to Moore's law today's electronic devices already belong to 'nanoelectronics', meaning that the individual devices have lateral dimensions smaller than 100 nm. At such small scales quantum effects become important. On the one hand, this means that classical electronic devices cannot be infinitely shrinked. On the other hand, these effects allow for a completely different kind of electronics at the nanoscale and possibly on-chip quantum computation.
The way to proceed is far from clear and open to creativity. Research in this field has been inspired by theoretical and experimental progress. On the theory side the understanding of interaction effects in quantum impurities, the characterisation of topological phase transitions and the development of exact numerical approaches, to mention a few, have triggered new experimental research. On the experimental side new materials such as nanowires, graphene, topological insulators and carbon nanotubes allowed for numerous developments beyond the scope of silicon-based technology.
Despite their profoundly different properties the nanoelectronic devices can still be easily integrated into standard microelectronics. Therefore, even if we cannot build all electronic devices the functional principles of which are based on quantum effects and new materials, we can still rely on the advances that have already been made.
In this thesis we concentrate on charge transport through nanoelectronic devices involving nontrivial correlated electronic states as Cooper pairs or excitons. Such systems may seem very different and physically indeed they are. However, from a theoretical perspective excitons may form a Bardeen-Cooper-Schrieffer condensate as electrons in superconductors and allow for similar theoretical treatments. These systems are of special importance since Cooper pairs are electron pairs in a singlet state. Splitting a singlet pair leads to a maximally entangled state of two electrons being an invaluable resource for quantum computing. Likewise, it has been demonstrated recently that superconductor hybrids can host Majorana fermions which can be used for topological quantum computation. Despite the abovementioned similarities of excitons and electron pairs there are also crucial differences allowing for several applications of exciton condensates, which are impossible to

realize in superconductors.

We will start this work with Chapter 2 providing the background of the different physical phenomena, which we will discuss in this thesis. We will illustrate the concept of full counting statistics, that provides access to all cumulants of the current flow including the noise. In Chapter 3 we will approach the simplest nanoelectronic devices involving a superconductor: quantum point contacts to normal conductors and ferromagnets. Chapter 4 will be concerned with the generation and detection of Majorana fermions in ferromagnetic wires and exciton condensates. Charge transport through the latter will be discussed in more detail in Chapter 5. Apart from the condensate phase of excitons we will also discuss how to create and detect a possible crystal phase.
Afterwards, we will move on to superconductor-hybrids incorporating quantum impurities. In Chapter 6 we will discuss in detail setups involving a single quantum dot and illustrate the principle of Cooper pair splitting. Especially, we will study the effect of onsite Coulomb repulsion due to the small dimensions of the quantum dot. Cooper pair splitters allow for the generation of entangled electron pairs and experimental studies have recently been carried out involving several double quantum dot setups. Their generic description will be developed in Chapter 7. A different class of superconductor-hybrids are junctions to a ferromagnet via a quantum dot which have recently been realized for the first time. We investigate such devices in Chapter 8 and illustrate the possible applications. We will discuss the effects of electron-phonon interaction in a superconductor-normal metal junction in Chapter 9 providing first steps towards the understanding of the interplay of vibrational degrees of freedom and the superconducting density of states. Finally, Chapter 10 will be devoted to the proposal of an experiment to witness the presence of entanglement in superconducting beamsplitters. We close this thesis in Chapter 11 by a discussion of the results and an outlook.

# Chapter 2

## Background

In this thesis we will discuss electronic transport through nanometer sized electronic devices involving the possibility of strong correlations between the electrons in the contacts as well as in the electronic device. Due to the small dimensions of the devices a classical treatment of the problems is futile and a quantum theoretical approach is required. In this Chapter we introduce the basic field-theoretical framework and the phenomena of superconductivity, the Kondo effect, exciton condensation and the generation of Majorana fermions. These will be the basic phenomena we will concentrate on in this work. Theoretical progress has been, to a large extent, accompanied and triggered by recent experimental advances both in the field of new semiconducting InAs (Csonka et al. [ 2008]) or InSb (Plissard et al. [ 2012]) nanowires and the realisation of exciton condensates both in semiconductors (Nandi et al. [ 2012]) and graphene (Gorbachev et al. [ 2012]) samples. The subsequent Sections summarize these basic concepts and give an overview of recent experimental progress.

### 2.1. Quantum transport theory and full counting statistics

A standard experimental method of nanoelectronics is the measurement of electric current through the mesoscopic sample which is connected to two reservoirs (source and drain) kept at a chemical potential difference $eV$. Often the device can already be sufficiently characterised by measuring $I(V)$ but a closer inspection reveals that the current shows specific fluctuations around its mean value. For macroscopic systems these fluctuations are Gaussian and therefore the probability $P(Q)$ for transferring a certain charge $Q$ through the device during a given measurement time $\tau$ can be characterised by its mean $\langle\langle Q\rangle\rangle$ and its variance $\langle\langle Q^2\rangle\rangle$. $\langle\langle Q^2\rangle\rangle$ is referred to as the noise and is non-vanishing even at finite temperature $T$ and zero $V$ (Johnson-Nyquist noise (Johnson [ 1928], Nyquist [ 1928])) and zero$T$ and finite $V$ (shot noise (Schottky [ 1918])).
This simple picture changes dramatically in the mesoscopic case. Even for the simplest system - a tunnel contact between two metals - the charge transfer statistics $P(Q)$ is a binomial distribution at $T = 0$ and a trinomial distribution at finite $T$ (Levitov and Lesovik [ 1994]). These distribution functions, of course, cannot be defined by their mean and variance alone and therefore $P(Q)$ deserves

closer inspection. However, from a theoretical point of view, the more suitable quantity is the moment generating function

$$\chi(\lambda) = \sum_{Q} e^{i\lambda Q} P(Q), \tag{2.1}$$

where $\lambda$ is referred to as the counting field. The logarithm of this quantity is the cumulant generating function providing direct access to the cumulants (irreducible moments) of the probability distribution,

$$\langle\langle Q^n \rangle\rangle = (-i)^n \left. \frac{\partial^n}{\partial \lambda^n} \right|_{\lambda=0} \ln \chi(\lambda). \tag{2.2}$$

We will concentrate on the limit of long measurement times $\tau$. In this case the first and second cumulant are associated with the time-averaged current and current noise via $\langle\langle Q \rangle\rangle = \tau I$ and $\langle\langle Q^2 \rangle\rangle = \tau S$. Higher cumulants can also be deduced from $\ln \chi(\lambda)$ but their interpretation is less intuitive. The first three cumulants representing the mean, variance and skewness of the distribution are shown for the binomial distribution in Fig. 2.1 . A Gaussian distribution with the same mean is shown in blue to illustrate the skewness.

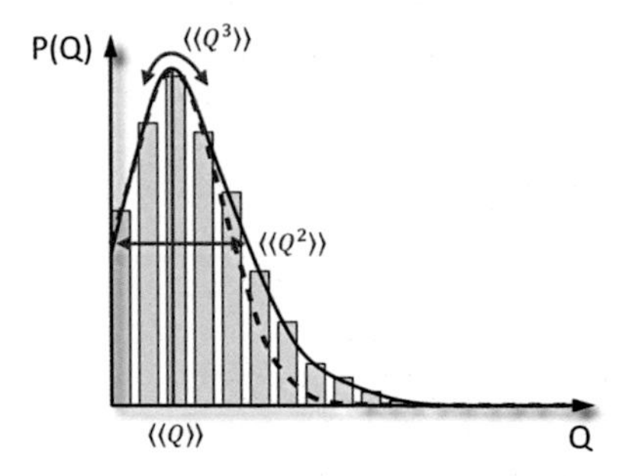


**Figure 2.1.:** Binomial charge transfer statistics $P(Q)$ (black) compared to a Gaussian distribution (blue): the red arrows indicate the first, second and third cumulant, referred to as mean, variance and skewness.

Next, we describe the method used to obtain $\chi(\lambda)$ theoretically. The calculation of higher order correlation functions requires special care with respect to time-ordering (Levitov and Lesovik [ 1994]). Manifold schemes have by now been developed to overcome this issue when calculating the cumulant generating function either based on path integrals (Nazarov and Kindermann [ 2003]) or the introduction of a tunneling operator (Maier [ 2011]). Both approaches are equivalent (Soller [ 2009]) and can equally well be implemented into the Keldysh formalism to calculate the full counting statistics (Nazarov [ 1999]). Here, a generalisation of this method designed for quantum impurity models (Gogolin and Komnik [ 2006b]) will be used.
The Keldysh formalism presented in Appendix A.1 is required as calculating averages in non-equilibrium systems is a non-trivial task. It assumes the system to be in equilibrium at time $t = -\infty$ and to introduce non-equilibrium adiabatically from $t = -\infty$ to $t_0 = 0$. Therefore, the only known eigenstates of the system exist at $t = -\infty$. From $t_0 = 0$ onwards we let the system evolve until some time $t$ and then go back in time to $t = -\infty$. The price to pay is that integrals for the expectation values need to be performed along the Keldysh contour shown in Fig. 2.2 .

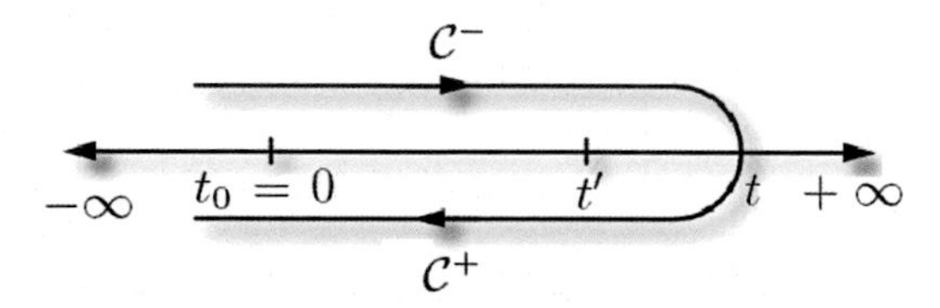


**Figure 2.2.:** Keldysh contour with the forward ($\mathcal{C}^-$) and backward ($\mathcal{C}^+$) parts. The interaction is introduced adiabatically until $t_0 = 0$ and then we let the system evolve until $t$ and return to $-\infty$.

The calculation of the cumulant generating function now proceeds via a modification of the tunnel Hamiltonian. We assume to have two contacts described by electron field operators $\Psi_R$ and $\Psi_L$ contacted at some point $x = 0$. The generic tunnel Hamiltonian is given by (Cohen et al. [ 1962])

$$H_T = \gamma[\Psi_R^+(x=0,t)\Psi_L(x=0,t) + \text{H.c.}], \tag{2.3}$$

and is applicable for any electrodes: metallic, ferromagnetic as well as superconducting. $t$ refers to the time of the tunnel event. In order to include information about the charge transferred through the device we need to include the counting field $\lambda$ introduced in Eq. (2.1). This can be done via the substitution $\Psi_R(x=0) \rightarrow \Psi_R(x=0)e^{-i\lambda(t)/2}$. The dependence $\lambda(t)$ was introduced on purpose. First, we want to extract the charge transferred during the interval $[0,\tau]$. Second, just introducing a constant phase will not suffice to extract information about the transferred charge since the contribution of the phase will average out when combining the forward and backward time propagation on the Keldysh contour $\mathcal{C}$ in Fig. 2.2 . Therefore$\lambda(t)$ must be explicitely time- and contour-dependent. It turns out that the appropriate choice is (Nazarov and Kindermann [ 2003])

$$\lambda(t) = \begin{cases} 0, & \text{for } t < 0, \\ \lambda, & \text{for } 0 < t < \tau \text{ and } t \in \mathcal{C}_-, \\ -\lambda, & \text{for } 0 < t < \tau \text{ and } t \in \mathcal{C}_+, \\ 0 & \text{for } t > \tau. \end{cases} \tag{2.4}$$

The tunnel operator after the substitution in Eq. (2.3) is given by

$$T^{\lambda(t)} = \gamma[\Psi_R^+(x=0,t)\Psi_L(x=0,t)e^{i\lambda(t)/2} + \Psi_L^+(x=0,t)\Psi_R(x=0,t)e^{-i\lambda(t)/2}]. \tag{2.5}$$

The interpretation of this expression is easy: an electron transferred from left to right is associated to $+\lambda$ and an electron transferred in the opposite direction is associated to $-\lambda$. The factor 1/2 is due to the separate evolution on the forward and backward branch of the time contour. The explicit contour-dependence of $\lambda(t)$ leads to an important difference compared to the traditional Keldysh formalism: the partition function is not unity anymore but depends on $\lambda$ and indeed this generalized partition function gives directly the moment-generating function defined in Eq. (2.1)

$$\chi(\lambda) = \left\langle T_{\mathcal{C}} \exp\left(-i\int_{\mathcal{C}} T^{\lambda(t)} dt\right)\right\rangle, \tag{2.6}$$

where $\mathcal{C}$ represents the Keldysh contour and $T_{\mathcal{C}}$ is the corresponding time-ordering.
The cumulant generating function as well as its cumulants represent measurable quantities, e.g. the third cumulant was measured in Reulet et al. [ 2003]. Several possible detection schemes for measuring the cumulant generating function have been put forward. Here we shall shortly describe

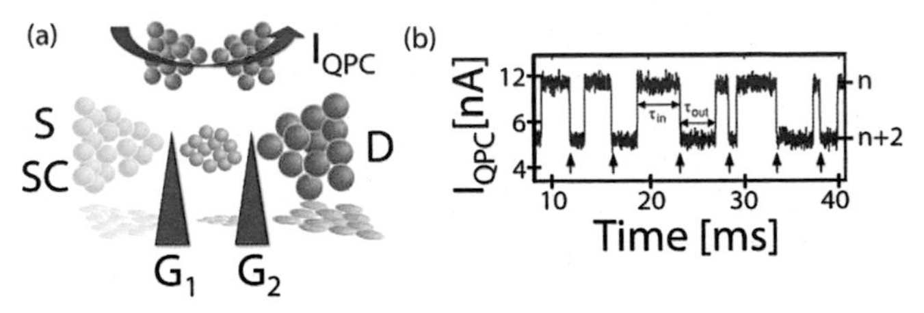


**Figure 2.3.: (a)**: sketch of the experimental setup for measuring the cumulant generating function. A quantum dot is contacted by a source ($S$) and drain ($D$) contact. $S$ in this case is a superconductor. The tunnel barriers can be controlled by two gates $G_1$ and $G_2$. The current through the upper quantum point contact can be measured as a function of time.
**(b)**: showcase time trace for the current through the quantum point contact.

the measurement scheme used in Gustavsson et al. [ 2007] for normal electrodes and sketched in Fig. 2.3.
A quantum dot is contacted by a source and a drain. These can be of different nature, e.g. the source could be a superconductor. Local shaping gates allow to tune the rates for electron transport between the different parts of the device. The quantum dot in turn is electrostatically coupled to a nearby quantum point contact. Due to this coupling, the properties of the quantum dot, i.e. its population, in turn affect the transmission characteristics of the quantum point contact. Hence, the time trace of the current through the quantum point contact reveals whether the dot is doubly occupied or not. In the case of a superconducting source only electron pairs on the dot are allowed due to the formation of Cooper pairs.
From such data sets recorded over a given measurement time $\tau$ we can count the charge transferred through the device and can even reveal the transition times $\tau_{\text{in}}$ and $\tau_{\text{out}}$ of individual electrons. This way one can directly access $P(Q)$ and consequently the cumulant generating function.

## 2.2. Superconductivity

Having mentioned Cooper pairs we should give a basic introduction to the main aspects of superconductivity, since we will investigate the abovementioned superconductor hybrid junctions later in closer detail. In spite of interesting aspects of exotic and high-$T_c$ superconductors we will restrict ourselves to classical superconductors described by the Bardeen-Cooper-Schrieffer (BCS) theory (Bardeen et al. [ 1957]) in the following. The mechanism leading to dissipationless current through superconductors is the formation of Cooper pairs which are bound states of two electrons correlated in momentum space. They form due to electron-phonon interaction which leads to the Cooper instability at low enough temperatures. This instability leads to a new ground state of the system with all electrons within an energy gap $\pm\Delta$ around the Fermi energy condensed into Cooper pairs. The formation of Cooper pairs allows for creation and annihilation of electron pairs out of the BCS condensate of Cooper pairs which in turn gives rise to finite expectation values of the kind $\langle c^+_{\mathbf{k}\uparrow} c^+_{-\mathbf{k}\downarrow}\rangle$. $c_{\mathbf{k}\sigma}$ refers to an annihilation operator for an electron with spin $\sigma$. The electrons forming a Cooper pair need to have opposite spin due to Fermi's exclusion principle. We will use the mean-field BCS

Hamiltonian in the form

$$H_S = \sum_{\mathbf{k}\sigma} \epsilon_{\mathbf{k}} c^+_{\mathbf{k}\sigma} c_{\mathbf{k}\sigma} - \Delta \sum_{\mathbf{k}} (c^+_{\mathbf{k}\uparrow} c^+_{-\mathbf{k}\downarrow} + c_{-\mathbf{k}\downarrow} c_{\mathbf{k}\uparrow}), \tag{2.7}$$

where $\epsilon_{\mathbf{k}}$ describes the kinetic energy of the electrons. Due to rotational symmetry the energy of the electrons only depends on $k = |\mathbf{k}|$. For a derivation from electron-phonon interaction and a discussion of its validity for all cases which we want to consider in the following we refer to Soller [2009]. The formation of Cooper pairs around the Fermi edge leads to a diverging superconducting density of states as a function of energy $\omega$ given by $\rho_S = \rho_{0S}|\omega|/\sqrt{\omega^2 - \Delta^2}$. $\rho_{0S}$ is constant in the wide band limit. Since the electrons in a Cooper pair form singlet states we also refer to Bardeen-Cooper-Schrieffer superconductivity as *s*-wave superconductivity and we can write the wavefunction of a Cooper pair as

$$\Psi_{\text{Cooper pair}} = \frac{1}{\sqrt{2}}(|\uparrow\rangle|\downarrow\rangle - |\downarrow\rangle|\uparrow\rangle), \tag{2.8}$$

if we neglect the momentum component of the electron field operators and just concentrate on the spin component.

## 2.3. InAs/InSb nanowires

The mesoscopic devices that fulfill the abovementioned size requirements and are employed in the contemporary experiments in question for this thesis are either based on carbon nanotubes or semiconducting nanowires. Carbon nanotubes have been the most popular system for the study of low-dimensional devices contacted to superconductors for more than a decade because of their low density of defects and the ability to grow them via chemical vapor deposition directly on a $SiO_2$ substrate (Herrmann et al. [ 2010]).
However, in the last few years semiconducting nanowires have attracted increasing interest. InAs and InSb are III-V semiconductor materials that have a large bulk Landé *g*-factor of up to 20 for InAs (Csonka et al. [ 2008]) or up to 80 for InSb (Plissard et al. [ 2012]) and large spin-orbit interaction. Semiconducting wires have the advantage that the carrier mobility can be tuned by gate voltages. InAs has the additional advantage that no Schottky barrier is formed in contact with a metal due to pinning of the metal's Fermi level in the conduction band. Experimentally it is found that such nanowires allow for good contact to superconductors (Mourik et al. [ 2012]) and even ferromagnets (Hofstetter et al. [ 2010]) which makes them one of the most promising building blocks of nanoelectronics.
Producing defect-free nanowires has been a big challenge in the past years since the bulk materials form a zincblende lattice whereas the wire geometry favors wurtzite structure due to its lower surface energy. This leads to defects of faulty stacking for the typical wire diameters of $\approx 150$ nm used in experiment. By now it has become possible to control the growth stage in order to produce (almost) defect-free wires (Johansson et al. [ 2010]).
When the nanowires are contacted by source and drain electrodes at small distance (typically around 300 nm) barriers are created close to the contacts if the carrier density is reduced by the appropriate gate voltages. This leads to the formation of small electronic islands known as quantum dots. Quantum mechanically they represent small potential boxes with a large but roughly constant energy spacing $\delta E$ for the spin-degenerate electronic levels. The capacitance $C$ of such small islands can be very small so that the occupation of both spin levels of the same energy level needs the large energy cost for adding one electron (charging energy) $U = e^2/C$. Typically the experimental situation is

such that $\delta E > U > k_B T, V$. Capacitive coupling of a gate electrode at a certain voltage $V_g$ to the quantum dot allows to vary the electronic levels on the quantum dot.
For sufficiently opaque tunnel barriers and zero source drain voltage $V$ we therefore expect conductance through a contact between two normal metals via a quantum dot only if the gate voltage is chosen such that an electronic level is aligned with the Fermi level of both electrodes, see Fig. 2.4 (a). In all other cases the current is blocked, see Fig. 2.4 (b).

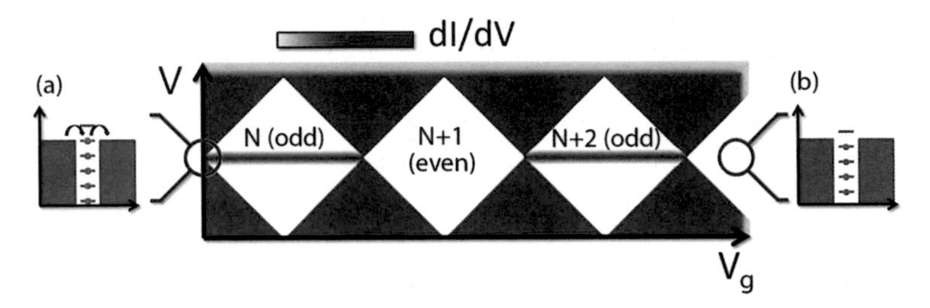


**Figure 2.4.:** Coulomb blockade: the electronic transport at zero bias $V$ is blocked (inset (b)) if an electronic level between source and drain is not on resonance (inset (a)). The electron number $N$ on the quantum dot is varied by the gate voltage $V_g$. At finite bias the diamonds close linearly. An additional resonance at zero bias appears for odd $N$.

If a finite bias voltage $V$ is applied the electrons can tunnel for all $V_g$ for which the electron level is in between the two Fermi levels of the electrodes. Therefore the current blocking is alleviated linearly leading to the typical Coulomb diamonds observable in Fig. 2.4 .

## 2.4. Kondo effect

The Kondo effect was first observed in disordered metals containing a finite number of magnetic impurities: the resistance below a critical temperature increased logarithmically with lowering $T$ (de Haas and van den Berg [ 1936]). These magnetic impurities are antiferromagnetically screened leading to the formation of a Kondo screening cloud of the conduction electrons which form a singlet state with the magnetic impurity (Kondo [ 1964]). This Kondo singlet is a much more effective scatterer than the impurity itself leading to a heavily increased scattering amplitude for conduction electrons at the Fermi edge for low temperatures.
This situation is reversed for quantum dot geometries. A quantum dot with an odd number of electrons will naturally act as a magnetic spin 1/2 impurity. However, in this case electron transport is only due to scattering so that the Kondo effect leads to an increased and even perfect conductivity which has been observed in many experiments (Goldhaber-Gordon et al. [ 1998], Kouwenhoven and Glazman [ 2001]). At low temperatures and large$\delta E$ the emerging zero-bias conductance peak observable in Fig. 2.4 is characterised by its width$T_K$, known as the Kondo temperature. In normal conducting systems the Kondo temperature is directly related to the onsite interaction $U$ via (Tsvelick and Wiegmann [ 1983])

$$T_K = \frac{\sqrt{2U\Gamma_1}}{\pi} \exp\left(-\frac{\pi U}{8\Gamma_1}\right), \tag{2.9}$$

where $\Gamma_1$ refers to the tunnel rate between the normal conductor and the quantum dot. At the strong coupling fixed point the perfectly transmitting channel via the Kondo resonance can be effectively described as a resonant level system as far as electronic transport is concerned (Ng and Lee [ 1988], Nozières [ 1974]).

## 2.5. Exciton condensation

An exciton is a bound pair of an electron and a vacancy (hole) that may occur in various circumstances. It is an elementary excitation e.g. accessible via laser-light and may be described by a quasiparticle in a way very similar to a Cooper pair. Possible realisations include $CuO_2$ in which radiative recombination of the excitons from the lowest exciton state is forbidden by symmetry (Snoke [2002]) and coupled two-dimensional electron gases (Eisenstein and MacDonald [ 2004]). The latter can be realised when an electron layer and a hole layer are separated by an insulating barrier that is sufficiently thick to prevent interlayer tunneling but sufficiently thin to induce interlayer Coulomb interaction. The two realisations of excitons differ: in bilayer quantum systems excitons can be created with the electron and the hole in the same layer (direct excitons) or in different layers (indirect excitons). In bulk semiconductors, of course, only direct excitons are possible. The massively increased lifetime of indirect excitons in layered semiconductor structures and the possible confinement of the electrons in those structures (Gärtner et al. [ 2007]) has triggered interest in this specific realisation of excitons.
Excitons are hydrogen-like Bose particles and may therefore condense. The type of condensation depends on the density (C. Comte and P. Nozières [ 1982]): for low densities a Bose Einstein condensate (BEC) is observed (High et al. [ 2012]) whereas for high densities exciton condensates are assumed to be BCS condensates. If particles Bose condense they become tightly bound whereas for a BCS condensate they are localised in momentum space (Zwierlein et al. [ 2004]). All important features in the BCS condensed situation are captured by an effectively one-dimensional model (as in Section 2.2 ) for an exciton condensate extending from$x = -l/2$ to $l/2$ (Dolcini et al. [ 2010])

$$H_{EC} = \int_{-l/2}^{l/2} dx\, \Psi^+(x) \begin{pmatrix} H_T & \Delta \\ \Delta & H_B \end{pmatrix} \Psi(x), \tag{2.10}$$

where $\Psi = (\Psi_T, \Psi_B)^T$ is the two-layer spinor and $H_T$ ($H_B$) describes the electron (hole) single-particle term of the top (bottom) layer. The interlayer pairing interaction is described by an exciton order parameter $\Delta$.
A remarkable advance in the field is expected to arise from graphene bilayers. Such exciton condensates are predicted to exhibit substantially higher critical temperature than ordinary semiconductor realisations due to the weaker screening and the higher electron and hole densities that can be achieved in graphene (Min et al. [ 2008]).
Whereas the BEC condensate phase has by now been observed experimentally via a Mach-Zehnder interferometer measurement (High et al. [ 2012]) the BCS phase has not been observed yet. However, the high density of excitons in this phase allows for an observation via the coherent motion of electrons and holes using an applied bias in one layer (Su and MacDonald [ 2008]). Recently, systems of two graphene layers separated by boron nitride have been realized (Gorbachev et al. [ 2012]) and substantial Coulomb drag even at room temperature has been observed.
Fig. 2.5 shows the measurement setup from ( Gorbachev et al. [ 2012]) where two graphene layers are embedded in a boron nitride crystal. The carrier densities in the top and bottom layer can be tuned by top gates at chemical potential $\mu_{T,B}$. Drag measurements were performed by applying a

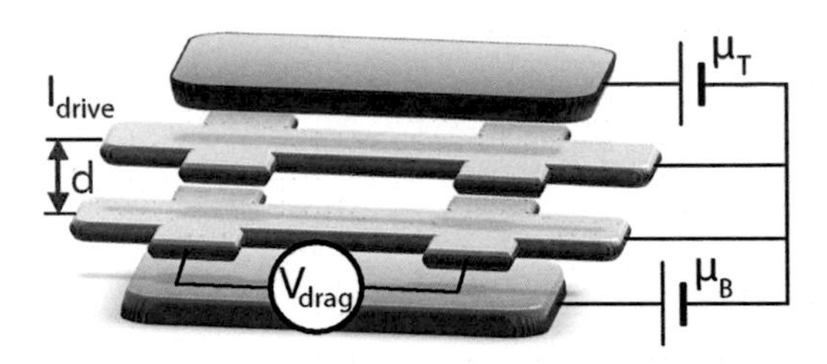


**Figure 2.5.:** Schematics of the measurement of drag current as a signature of the BCS condensate phase of excitons: Two graphene layers are placed at a distance $d$ in a boron nitride crystal. Carrier densities in the two graphene layers can be tuned by applying different chemical potentials $\mu_T$ and $\mu_B$. A drive current can be injected in the top layer and the resulting voltage drop in the lower layer $V_{\mathrm{drag}}$ is measured.

drive current $I_{\mathrm{drive}}$ in one layer and measuring the induced voltage in the second layer.
Another interesting feature of excitons is the excistence of parallel electric dipole moments of indirect excitons, which become important at intermediate densities. In this case the repulsive interaction leads to an emergent long-range order of Wigner-crystal type (Astrakharchik et al. [ 2007]).

## 2.6. Majorana fermions

A Majorana fermion is a fermion which is its own antiparticle. Consequently, a Majorana fermion field $\gamma_A$ with a property $A$ at position $x$ fulfills $\gamma_{A,x} = \gamma^{+}_{A,x}$. Recently, it has been proposed that Majorana fermions can be realized in a solid state environment (Kitaev [ 2001]).
The author proposed to place a quantum wire on top of a spin-polarized $p$-wave superconductor. Spin-polarized $p$-wave superconductors are superconductors with both electrons in a Cooper pair having equal spin compared to opposite spin in Eq. (2.7). The superconductor would induce $p$-wave correlations in the quantum wire described by an order parameter $\Delta_p$ so that the tight binding Hamiltonian for the wire of finite length at chemical potential $\mu_{\mathrm{wire}}$ can be written as

$$H = -\mu_{\mathrm{wire}} \sum_{x=1}^{N} c_x^{+} c_x + \sum_{x=1}^{N-1} (-t_{\mathrm{wire}} c_x^{+} c_{x+1} + \Delta_p c_x c_{x+1} + \mathrm{H.c.}), \tag{2.11}$$

where $t_{\mathrm{wire}}$ refers to the local tunneling strength. For simplicity let us consider the case $\mu_{\mathrm{wire}} = 0, t_{\mathrm{wire}} = \Delta_p$ in which we can write

$$H = \sum_{x=1}^{N-1} (-t_{\mathrm{wire}} c_x^{+} c_{x+1} + t_{\mathrm{wire}} c_x c_{x+1} + \mathrm{H.c.}). \tag{2.12}$$

Here, the fermionic operators can be written in terms of Majorana operators $\gamma_{\alpha,x} = \gamma^{+}_{\alpha,x}$, $\alpha = A,\ B$ via

$$c_x = \frac{1}{2}(\gamma_{B,x} + i\gamma_{A,x}),\ c_x^{+} = \frac{1}{2}(\gamma_{B,x} - i\gamma_{A,x}). \tag{2.13}$$

Then from Eq. (2.12) we find

$$H = -it_{\mathrm{wire}} \sum_{x=1}^{N-1} \gamma_{B,x}\gamma_{A,x+1}, \tag{2.14}$$

which is a simple tight-binding Hamiltonian for Majorana fermions. A similar mapping can be achieved at $\Delta_p \neq t_{\text{wire}}$, which, however, is more complicated. Two of the Majoranas always combine into an ordinary fermion $d_x = (\gamma_{A,x+1} + i\gamma_{B,x})/2$. The formation of a fermion therefore costs $2\Delta_p = 2t_{\text{wire}}$ of energy so that the Majorana fermion picture survives up to $|\mu_{\text{wire}}| = 2t_{\text{wire}}$.
In this formulation two Majorana fermions are uncoupled: $\gamma_{A,1}$ and $\gamma_{B,N}$. These states are bound states at zero energy localized at both ends of the wire, which are uncoupled from the other states in the wire, see Fig. 2.6 . They emerge in superconductors as the quasiparticles that can be excited at zero energy are symmetric superpositions of electrons and holes (see also Appendix A Eq. ( A.17)).

**Figure 2.6.:** Sketch of the experimental device for creating a pair of Majorana fermions in an InSb nanowire in proximity to an $s$-wave superconductor. An external **B**-field is applied parallel to the wire whereas the spin-orbit interaction of the wire induces an effective magnetic field $\mathbf{B}_{\text{so}}$ perpendicular to the wire. The position of the Majorana fermions is indicated by yellow spots.

Experimentally $p$-wave superconductors with the desired properties are hard to find. Therefore, finding systems that have the low-energy behavior described by the Hamiltonian in Eq. (2.11) has trigerred a lot of research activity (Gangadharaiah et al. [ 2011], Oreg et al. [ 2010]). Typically, the starting point is a one-dimensional nanowire made of InAs/InSb placed on a superconducting substrate. As described in Section 2.3 these nanowires exhibit strong spin-orbit coupling. The effect can be rationalized as having an effective magnetic field $\mathbf{B}_{\text{so}}$ perpendicular to the wire direction, which leads to spin-flips at the interface between the superconductor and the nanowire. In combination with the proximity coupled superconductor this leads to $p$-wave correlations in the nanowire. In the presence of an additional external magnetic field along the wire a gap in the spectrum is opened at the crossing between the two spin bands, which makes the nanowire effectively spin-polarized. In total this creates the same low-energy behavior as for the Hamiltonian in Eq. (2.11).
The Majorana bound states at the end of the wire can be accessed from the outside by additional electrodes. Their presence leads to a peak in the conductance at zero energy which is exactly the feature that has been observed experimentally (Mourik et al. [ 2012]).

# Chapter 3

## Superconducting tunnel contacts

The first class of systems we want to consider are superconducting tunnel contacts, either involving a normal metal or a ferromagnet as the second electrode. These systems belong to the best-studied in quantum transport (e.g. Blonder et al. [1982], Soulen et al. [1998]). In the following Sections we will first describe the normal-superconductor quantum point contact (QPC) in order to introduce the method and possible approximations. In the second part we describe the ferromagnet-superconductor QPC. These simple systems on the one hand represent ideal benchmarks for our method to calculate the cumulant generating function (CGF). On the other hand the availability of high-quality experimental data and exact theoretical predictions allows for a detailed check of the assumptions made in modelling the experiments (Xia et al. [2002]). In order to simplify notation from now on we use units $e = \hbar = k_B = 1$ and only restore SI units when necessary.

### 3.1. Normal-superconductor tunnel contact

The first system we want to consider is a QPC between a normal metal and a superconductor (SC). The CGF of charge transfer in this system has been calculated either using wave function matching (Muzykantskii and Khmelnitskii [1994]), circuit theory (Belzig [2002], Belzig and Nazarov [2001a]) or a Hamiltonian description (Soller and Komnik [2011a]) that will also be used here. The system is described by the following Hamiltonian

$$H_{\text{NSQPC}} = H_1 + H_S + H_{T1}, \tag{3.1}$$

where $H_1$ refers to the normal metal, which is modelled as a noninteracting fermionic continuum at chemical potential $\mu_1$, written in terms of electron field operators $\Psi_{k1\sigma}$ with a constant density of states (DOS) given by $\rho_{01}$. $H_S$ is given by Eq. (2.7) and the tunneling Hamiltonian $H_{T1}$ in the wide flat band limit is given by a simple extension of Eq. (2.3)

$$H_{T1} = \sum_\sigma \gamma_1 [\Psi^+_{1\sigma}(x=0,t) c_\sigma(x=0,t) + \text{H.c.}], \tag{3.2}$$

involving the tunnel amplitude $\gamma_1$. As in previous treatments of SC hybrids we keep the SC in equilibrium ($\mu_S = 0$), so that the applied voltage is given by $V = \mu_S - \mu_1 = -\mu_1$.
We assume to have a local contact only at one point $x = 0$ as already pointed out in Eq. (2.3). In this case the size and confining potential of this contact is assumed to be such that a strong focussing in the two dimensions transverse to the direction of current flow is observed. For simplicity the electron propagation is assumed to be in the $x$-direction. Due to the strong focussing by the confining potential $U_C(y,z)$ the wavefunctions of a charge carrier at position $\mathbf{r} = (x,y,z) = (x,\mathbf{R})$ can be written as a superposition [Datta, 1997, p. 11]

$$\Psi(\mathbf{r},t) = \sum_j \phi_j(\mathbf{R},t)\Psi_j(x,t), \tag{3.3}$$

where $\{\phi_j(\mathbf{R},t),\ j = 1,\cdots,N\}$ is a complete basis of eigenfunctions of the confining potential $U_C(y,z)$. Usually, at low temperatures, only the lowest mode $j = 1$ is occupied and the higher modes do not play any significant role. This is also known as $s$-wave scattering. Then the $y$- and $z$-dimension can both be neglected and the conductor may be treated as a one-dimensional system so that we do not have $\mathbf{k}$ in Eq. (2.7) anymore but scalars $k$.
We introduce the counting field via the same transformation leading to Eq. (2.5), so that we arrive at the transformed tunnel Hamiltonian

$$T_1^\lambda = \sum_\sigma \gamma_1[e^{-i\lambda/2}\Psi_{1\sigma}^+(x=0,t)c_\sigma(x=0,t) + \text{H.c.}]. \tag{3.4}$$

For the calculation of the CGF we follow the approach outlined in (Gogolin and Komnik [2006b]): for long measurement times $\tau$ the CGF may be written using an adiabatic potential $U_a(\lambda)$ (Hamann [1971]) (that by construction does not depend on time) as

$$\ln \chi_{\text{NSQPC}}(\lambda) = -i\int_0^\tau dt U_a(\lambda) = -i\tau U_a(\lambda). \tag{3.5}$$

In turn the adiabatic potential is related to the counting field derivative of $T_1^\lambda$ in Eq. (3.4). The time $t$ introduced in Eq. (3.2) is an ordinary time on the $\mathcal{C}^-$-branch of the Keldysh contour (see Fig. 2.2). Therefore the the counting field derivative also has to be performed with $\lambda$ on the $\mathcal{C}^-$-branch

$$\frac{\partial U_a}{\partial\lambda^-} = \left\langle T_\mathcal{C}\frac{\partial T_1^\lambda}{\partial\lambda^-}\right\rangle_\lambda = -\frac{i\gamma}{2}\sum_\sigma\langle T_\mathcal{C}\Psi_{1\sigma}^+(x=0,t)c_\sigma(x=0,t)\rangle_\lambda e^{-i\lambda/2} + \text{H.c.}, \tag{3.6}$$

where $\langle\,\cdot\,\rangle_\lambda$ is defined as $\langle\,\cdot\,\rangle_\lambda := 1/\chi(\lambda)\,\langle\,\cdot\,\rangle$ with $\langle\,\cdot\,\rangle$ being the ordinary expectation value with respect to the system's Hamiltonian in Eq. (3.1) with $H_{T1}$ replaced by $T_1^\lambda$.
In order to simplify the calculation we introduce (inhomogeneous) Green's functions (GFs)

$$\mathcal{G}_{1S\sigma}^\lambda(t,t') = i\langle T_\mathcal{C}\Psi_{1\sigma}^+(x=0,t')c_\sigma(x=0,t)\rangle_\lambda,\ \mathcal{G}_{S1\sigma}^\lambda(t,t') = i\langle T_\mathcal{C}c_\sigma^+(x=0,t')\Psi_{1\sigma}(x=0,t)\rangle_\lambda.$$

In Eq. (3.6) the times $t,\ t'$ belong to the $\mathcal{C}^-$-branch of the Keldysh contour so that we only need the '$--$'-component of the GFs defined above. These inhomogeneous GFs may be calculated exactly since the Hamiltonian in Eq. (3.1) is quadratic. Taking into account the properties of the aforementioned $\lambda$-expectation values we obtain the following relations

$$\begin{aligned}
\mathcal{G}_{1S\sigma}^\lambda(t,t') &= \gamma_1\int_\mathcal{C} ds\left[e^{-i\lambda(s)/2}\mathcal{G}_{1\sigma}^\lambda(t,s)g_{S\sigma}(s,t')\right],\\
\mathcal{G}_{S1\sigma}^\lambda(t,t') &= \gamma_1\int_\mathcal{C} ds\left[e^{i\lambda(s)/2}g_{S\sigma}(t,s)\mathcal{G}_{1\sigma}^\lambda(s,t')\right],
\end{aligned}$$

where $g_{S\sigma}$ and $\mathcal{G}^{\lambda}_{1\sigma}$ refer to the bare SC GF and the exact-in-tunneling and $\lambda$-dependent GF of the normal metal. For a calculation of the bare SC GF we refer to Appendix A Eq. ( A.23). Fourier transformation leads to

$$\left\langle T_C \frac{\partial T_1^{\lambda}}{\partial \lambda^-} \right\rangle = -\frac{\gamma_1^2}{4} \sum_{\sigma} \int \frac{d\omega}{2\pi} \left[ e^{i\lambda} g_{S\sigma}^{+-}(\omega) \mathcal{G}_{1\sigma}^{\lambda-+}(\omega) - e^{-i\lambda} g_{S\sigma}^{-+}(\omega) \mathcal{G}_{1\sigma}^{\lambda+-}(\omega) \right]. \tag{3.7}$$

Therefore, we have reduced the problem to the calculation of the exact-in-tunneling and $\lambda$-dependent GF of the normal metal. This is done by means of the Dyson equation (Cuevas et al. [ 1996]). The SC induces superconducting correlations also in the normal metal. Therefore we have to introduce a superconducting order parameter also at the interface of the normal metal. We do so by introducing anomalous GFs as $\mathcal{F}^{\lambda}_{kk'1}(t,t') = i\langle T_C \Psi_{-k1\downarrow}(t) \Psi_{k'1\uparrow}(t')\rangle_{\lambda}$, $(\mathcal{F}^{\lambda}_{kk'1})^{+}(t,t') = i\langle T_C \Psi^{+}_{k1\uparrow}(t) \Psi^{+}_{-k'1\downarrow}(t')\rangle_{\lambda}$. Additionally, we introduce the $\lambda$-dependent self energies $\Sigma_S^n$ and $\Sigma_S^a$ referring to the normal and anomalous contributions by the superconducting drain as

$$\begin{aligned} \Sigma_S^n(\omega) &= \gamma_1^2 \begin{pmatrix} g_S^{--}(\omega) & -e^{-i\lambda} g_S^{-+}(\omega) \\ -e^{i\lambda} g_S^{+-}(\omega) & g_S^{++}(\omega) \end{pmatrix}, & (3.8) \\ \Sigma_S^a(\omega) &= \gamma_1^2 \begin{pmatrix} f_S^{--}(\omega) & -e^{-i\lambda} f_S^{-+}(\omega) \\ -e^{i\lambda} f_S^{+-}(\omega) & f_S^{++}(\omega) \end{pmatrix}, & (3.9) \end{aligned}$$

where $f_S(\omega)$ refers to the anomalous GF in the SC as calculated in Eq. (A.23). Combining the Dyson equations for both GFs in the normal electrode leads to

$$\begin{pmatrix} \mathcal{G}^{\lambda}_{1\sigma}(\omega) \\ \mathcal{F}^{\lambda}_{1}(\omega) \end{pmatrix} = \left( \begin{matrix} g_1(\omega) \\ \begin{pmatrix} 0 & 0 \\ 0 & 0 \end{pmatrix} \end{matrix} \right) \left[ \mathbf{1} - \begin{pmatrix} \Sigma_S^n(\omega) g_{1\sigma}(\omega) & \Sigma_S^a(\omega) g_{1\sigma}(\omega) \\ \Sigma_S^a(\omega) g_{1-\sigma}(-\omega) & \Sigma_S^n(\omega) g_{1-\sigma}(-\omega) \end{pmatrix} \right]^{-1}, \tag{3.10}$$

where the notation of the second matrix refers to a $2 \times 4$-matrix consisting of $g_1(\omega)$ and further zero-entries. In this way we obtain an exact expression for the temperature- and energy-dependent CGF that is valid over the whole range of possible parameters. It is given by

$$\begin{aligned} &\ln \chi_{\mathrm{NSQPC}}(\lambda, \tau) = \\ &\tau \int \frac{d\omega}{\pi} \Bigg[ \ln \Bigg( \prod_{\alpha=\pm} \left\{ 1 + T_e(\omega) \left[ n_{1\alpha}(1-n_S)(e^{i\alpha\lambda}-1) + n_S(1-n_{1\alpha})(e^{-i\alpha\lambda}-1) \right] \right\} \\ &+ T_{A2}(\omega)(2n_S-1) \Big\{ (2n_S-1) \Big[ (e^{i\lambda}-1)^2 n_{1-}(1-n_{1+}) - 2(e^{i\lambda}-1)(e^{-i\lambda}-1) n_{1-} n_{1+} \\ &+ (e^{-i\lambda}-1)^2 n_{1+}(1-n_{1-}) \Big] + 2n_S(e^{i\lambda}-1)(e^{-i\lambda}-1)(n_{1+}-1+n_{1-}) \Big\} \\ &+ T_{\mathrm{BC}}(\omega)(2n_S-1)(e^{i\lambda}-e^{-i\lambda})^2 \Big\{ (2n_S-1)[n_{1-}e^{i\lambda} + n_{1+}e^{-i\lambda} + \Gamma_e n_S(1-n_S)(e^{i\lambda}-e^{-i\lambda})^2 \\ &- (n_{1+}-1+n_{1-}) n_S (e^{i\lambda}+e^{-i\lambda})] - 4n_S(1-n_S)(n_{1+}-1+n_{1-}) \Big\} \theta\left( \frac{|\omega|-\Delta}{\Delta} \right) \Bigg) \\ &+ \ln \left\{ 1 + T_A(\omega) \left[ n_{1+}(1-n_{1-})(e^{2i\lambda}-1) + n_{1-}(1-n_{1+})(e^{-2i\lambda}-1) \right] \right\} \theta\left( \frac{\Delta-|\omega|}{\Delta} \right) \Bigg], \end{aligned} \tag{3.11}$$

where the effective transmission coefficients are given by

$$\begin{aligned} T_e(\omega) &= \frac{4\Gamma_e}{[(1+\Gamma_e)^2 - \Gamma_A^2]}, \; T_{A2}(\omega) = \frac{4\Gamma_A^2}{[(1+\Gamma_e)^2 - \Gamma_A^2]^2} = \frac{T_{\mathrm{BC}}(\omega)}{\Gamma_e} \text{ and} \\ T_A(\omega) &= \frac{4\Gamma_A^2}{\Gamma_A^4 + 2\Gamma_A^2(1-\Gamma_e^2) + (1+\Gamma_e^2)^2}. \end{aligned} \tag{3.12}$$

The energy-dependent DOS of the superconducting terminal affects the hybridisations given by $\Gamma_e = \Gamma|\omega|/\sqrt{|\omega^2 - \Delta^2|}$ and $\Gamma_A = \Gamma\Delta/\sqrt{|\Delta^2 - \omega^2|}$, where $\Gamma = \pi^2 \rho_{01}\rho_{0S}\gamma_1^2/2$. The subscripts $e$ and $A$ refer to the electronic and Cooper pair DOS. $T_e$ refers to the transfer of single electrons and $T_{A2}$, $T_{\mathrm{BC}}$ describe the additional noise contributions by so called Andreev reflection (AR) above the gap and branch-crossing processes, respectively. AR is the process of retro-reflection of an electron at the SC interface as a hole accompanied by a Cooper pair formation in the SC. Branch-crossing refers to an electron entering the SC and forming a Cooper pair leading to the formation of a hole. $T_A$ is the transmission coefficient of AR processes below the gap. The Fermi distribution for the normal/SC lead is abbreviated by $n_{1+}/n_S$. $n_{1-} = 1 - n_{1+}(-\omega)$ refers to hole-like contributions.
For $\Delta \to 0$ only the first line of the expression in Eq. (3.11) remains which is the sum of electrons and holes of the expression for the CGF of a normal tunnel contact derived in Levitov and Lesovik [ 1994]. Consequently the dominating charge transfer events above the gap are single-electron transfers. Below the gap only terms proportional to $e^{2i\lambda}$ occur referring to double electron transfers in AR processes.
We calculate the current corresponding to this CGF via

$$I_{\mathrm{NSQPC}} = -\frac{i}{\tau}\frac{d}{d\lambda} \ln \chi_{\mathrm{NSQPC}}(\lambda, \tau)|_{\lambda=0}. \tag{3.13}$$

Considering the differential conductance $G_{\mathrm{NSQPC}} = dI_{\mathrm{NSQPC}}/dV$ for low voltages and $T = 0$ we obtain the result previously obtained in Beenakker [ 1992]. Likewise, the equivalence of the widely used Blonder-Tinkham-Klapwijk (BTK) model (Blonder et al. [ 1982]) and the Hamiltonian approach used here has been demonstrated in Cuevas et al. [ 1996]. We also reproduce exactly the results for the non-linear current voltage characteristics in Cuevas et al. [ 1996]. To further demonstrate the validity of the Hamiltonian approach we also performed a comparison to the experimental results for Al/Cu contacts with high transparency presented in Pérez-Willard et al. [ 2004], see Fig. 3.1 .

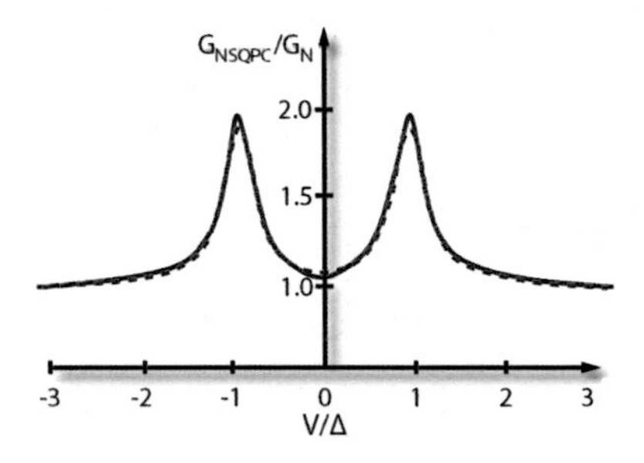


**Figure 3.1.:** Comparison of the conductance of an Al/Cu contact at $T = 95$ mK (red dashed curve) taken from Pérez-Willard et al. [ 2004] and the theoretical prediction using Eq. (3.13) with the fit parameter $\Gamma = 0.37$. The conductance is given in multiples of the normal state conductance $G_N$ that can be determined by setting $\Delta \to 0$.

Using the mapping to the BTK model in Eq. (3.11) we recover the results obtained in Belzig [ 2002], Belzig and Nazarov [ 2001a] and Muzykantskii and Khmelnitskii [ 1994].
The result in Eq. (3.11) is quite lengthy. Often approximations offer a better way of rationalizing the result. For low energy $\omega \ll \Delta$ only the anomalous GF in the SC will contribute and we arrive at

$$f_S(\omega) = \pi \frac{\Delta \rho_{0S}}{\sqrt{\Delta^2 - \omega^2}} \begin{pmatrix} 1 & 0 \\ 0 & -1 \end{pmatrix}, \; g_S(\omega) = 0. \tag{3.14}$$

In the limit of high energy $\omega \gg \Delta$ only the normal GF in the SC will contribute and we arrive at

$$f_S(\omega) = 0,\ g_S(\omega) = i\pi \frac{\omega \rho_{0S}}{\sqrt{\omega^2 - \Delta^2}} \begin{pmatrix} 2n_S - 1 & 2n_S \\ 2n_S - 2 & 2n_S - 1 \end{pmatrix}. \tag{3.15}$$

These approximations often allow for great simplifications in their regimes of applicability (Fazio and Raimondi [1998]). Using the simplified GFs of Eq. (3.14) for $|\omega| < \Delta$ and Eq. (3.15) for $|\omega| > \Delta$ in Eq. (3.7) we arrive at the CGF

$$\begin{aligned} &\ln \chi_{\text{NS, approx}} = \\ &2\tau \int \frac{d\omega}{\pi} \left( \ln\left\{1 + T_{\text{e,a}}(\omega)[n_{1+}(1-n_S)(e^{i\lambda}-1) + n_S(1-n_{1+})(e^{-i\lambda}-1)]\right\} \theta\left(\frac{|\omega|-\Delta}{\Delta}\right) \right. \\ &\left. + \frac{1}{2}\ln\left\{1 + T_{\text{A,a}}(\omega)\left[n_{1+}(1-n_{1-})(e^{2i\lambda}-1) + n_{1-}(1-n_{1+})(e^{-2i\lambda}-1)\right]\right\} \theta\left(\frac{\Delta-|\omega|}{\Delta}\right)\right), \end{aligned} \tag{3.16}$$

where the transmission coefficients are given by

$$T_{\text{e,a}}(\omega) = \frac{4\Gamma_e}{(1+\Gamma_e)^2} \text{ and } T_{\text{A,a}}(\omega) \quad = \quad \frac{4\Gamma_A^2}{(1+\Gamma_A^2)^2}. \tag{3.17}$$

Interestingly, the expression in Eq. (3.16) also gives qualitatively correct results for $|\omega| \approx \Delta$.
The above described approximation also works in the limit of low tunnel coupling $\gamma_1$ since the additional GFs only give rise to higher order corrections in the tunnel coupling in the respective energy regimes leading to AR and branch-crossing above the gap so that Eq. (3.16) is a correct CGF in the limit of a low-transparency tunnel contact.
Indeed, in the regime of low tunnel coupling we can be even more drastic and neglect AR altogether since also below the gap it is of higher order in the tunnel coupling and arrive at

$$\begin{aligned} &\ln \chi_{\text{NS, semi}} \\ = \quad &2\tau \int \frac{d\omega}{\pi} \ln\left\{1 + T_{\text{e,semi}}(\omega)[n_{1+}(1-n_S)(e^{i\lambda}-1) + n_S(1-n_{1+})(e^{-i\lambda}-1)]\right\} \theta\left(\frac{|\omega|-\Delta}{\Delta}\right), \end{aligned} \tag{3.18}$$

where $T_{\text{e,semi}}(\omega) = 4\Gamma_e$. This approximation is known as the semiconductor model (Tinkham [1996]). Calculating the differential conductance $G_{\text{NS, semi}}$ from Eq. (3.18) and comparing it with the result $G_{\text{N, semi}}$ when setting $\Delta = 0$ we arrive at the relation

$$\frac{G_{\text{NS, semi}}}{G_{\text{N, semi}}} = \rho_S(V). \tag{3.19}$$

Thus, the differential conductance measures directly the SC DOS. Using metals or ferromagnets for spectroscopy in the SC therefore has a long history (Giaever and Megerle [1961]).

## 3.2. Ferromagnet-superconductor tunnel contact

The second tunnel contact we want to investigate is one between a SC and a ferromagnet (FM). In this case Eq. (3.1) is modified by an exchange of the normal metal with a FM so that the Hamiltonian of the system reads

$$H = H_F + H_S + H_{T1}, \tag{3.20}$$

where, for the FM, we use electron field operators $\Psi_{kF\sigma}$ instead of $\Psi_{k1\sigma}$ in $H_{T1}$. $H_F$ describes the FM lead in the Stoner model with an exchange energy $h_{\text{ex}}$ as in Mélin [2004]

$$H_F = \sum_{k,\sigma} \epsilon_k \Psi^+_{kF\sigma}\Psi_{kF\sigma} - h_{\text{ex}} \sum_k (\Psi^+_{kF\uparrow}\Psi_{kF\uparrow} - \Psi^+_{kF\downarrow}\Psi_{kF\downarrow}). \tag{3.21}$$

The FM has a fermionic flat band DOS with asymmetry for the spin-$\uparrow$ and spin-$\downarrow$ tunneling electrons $\rho_{0F\sigma} = \rho_{0F}(1+\sigma P)$, where $P$ is the polarisation. Again we define the voltage such that $V = -\mu_F$, where $\mu_F$ is the chemical potential of the FM. $n_{F+} = 1 - n_{F-}(-\omega)$ refer to the electron- and hole-like Fermi distributions in the FM, respectively.
We use again Eq. (2.6) and follow the steps outlined in Section 3.1 in order to arrive at the CGF for the system described by the Hamiltonian in Eq. (3.20). Furthermore, we use the approximation for the SC GFs introduced in Eqs. (3.14) and (3.15) for simplification. The full result is given in Appendix B Eq. (B.1). The result involving the abovementioned approximations is

$$\begin{aligned}
&\ln \chi_{\text{SFQPC}}(\lambda) = \\
&2\tau \int \frac{d\omega}{2\pi} \left( \sum_\sigma \ln \left\{ 1 + T_{e\sigma,\text{a}}[n_{F+}(1-n_S)(e^{i\lambda}-1) + n_S(1-n_{F+})(e^{-i\lambda}-1)] \right\} \theta\left(\frac{|\omega|-\Delta}{\Delta}\right) \right. \\
&\left. + \ln \left\{ 1 + T_{\text{AF,a}}[n_{F+}(1-n_{F-})(e^{2i\lambda}-1) + n_{F-}(1-n_{F+})(e^{-2i\lambda}-1)] \right\} \theta\left(\frac{\Delta-|\omega|}{\Delta}\right) \right),
\end{aligned} \tag{3.22}$$

involving the effective transmission coefficients

$$T_{e\sigma,\text{a}} = \frac{4\Gamma_e(1+\sigma P)}{[1+\Gamma_e(1+\sigma P)]^2}, \text{ and } T_{\text{AF,a}} = \frac{4\Gamma_A^2(1-P^2)}{[1+\Gamma_A^2(1-P^2)]^2}. \tag{3.23}$$

We can easily recover Eq. (3.16) by choosing $P = 0$. The transmission coefficient $T_{e\sigma,\text{a}}$ refers to single-electron transfer while $T_{\text{AF,a}}$ is the transmission coefficient for AR processes below the gap. This demonstrates that also in the case of SC-FM-QPCs the elementary processes of charge transfer can be identified as normal electron transfer between the electrodes above the gap and AR processes below the gap, see Fig. 3.2 (c) and (a).
$T_{\text{AF,a}}$ has a strong dependence on the polarisation. Consequently, fitting the differential conductance of a SC-FM-QPC using the CGF in Eq. (B.1) allows for a precise determination of the polarisation of ferromagnets (Soulen et al. [1998]). However, if we compare the results we obtain for the differential conductance to the experimental results for Al/Co contacts with good transparencies (Pérez-Willard et al. [2004]) we need to introduce a sizeable broadening of the BCS DOS described by a Dynes parameter $\Gamma_D$ (Dynes et al. [1978]) that has to be of the order $\Gamma_D = 0.21\Delta$ to obtain quantitative agreement. Such a distortion is unexpected since the Al/Cu contacts fabricated by the same experimental procedure do not show any distorted BCS DOS, see Fig. 2.6.
This is not a problem of the Hamiltonian approach but has also been encountered when fitting $I - V$ spectra to an extension of the BTK model (Martín-Rodero et al. [2001]). Instead of a Dynes parameter one could introduce an effective temperature, but a reliable explanation of the spectra under debate (Xia et al. [2002]) may only be obtained by changing the model of the interface region (Grein et al. [2009]).

## 3.3. SC-FM tunnel contact with spin-active scattering

In the previous Section it is mentioned that a realistic description of the interface region is necessary for a complete understanding of SC-FM-QPCs. This can be achieved by considering a more

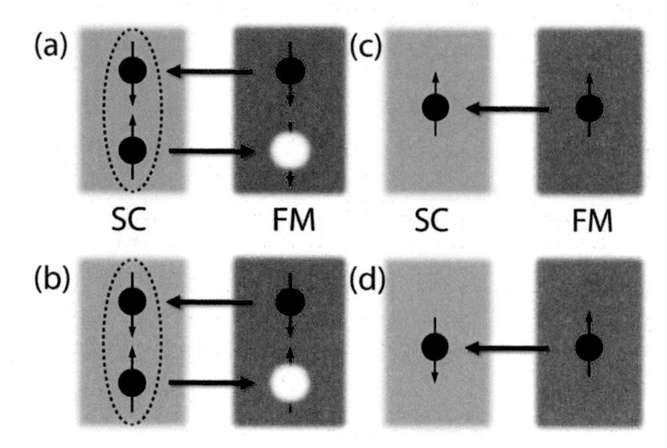


**Figure 3.2.:** Different transport processes in a SC-FM-QPC with spin-active scattering: in **(a)** we show the typical AR and in **(c)** we show the typical single-electron transmission between a SC and a FM. In **(b)** we show the spin-flipped Andreev process involving a spin-flip at the interface and giving rise to triplet correlations in the FM. Likewise also spin-flip transmissions occur as indicated in **(d)**.

complex model that explicitely includes a spin-dependent scattering potential (Grein et al. [ 2010]) at the interface. This is of special importance when dealing with the experimentally relevant case of strong spin polarisation $P \approx 0.2 - 0.8$. The mechanism of spin-active scattering at the interface is the interplay of the ferromagnetic exchange field in both the bulk and the interface. In the simplest case the two magnetic moments deep in the bulk and at the interface would just be parallel. However, manifold processes may lead to an interface magnetic moment different from the bulk like using a thin magnetic layer, spin-orbit coupling, magnetic anisotropy or spin-relaxation.
Possible theoretical treatments of point contact spectra include a quasi-classical Green's function approach (Cottet et al. [ 2009], Grein et al. [ 2010]), a wave-function matching technique (Duckheim and Brouwer [ 2011]) and a scattering states description (Cottet et al. [ 2008]). However, access to the CGF so far has been demonstrated only via a Hamiltonian approach (Soller et al. [ 2012d]) as in Section 3.2 .
The spin-active scattering is included via an additional spin-flip contribution to the tunnel Hamiltonian (Yamada et al. [ 2007])

$$H_{T2} = \sum_{\sigma} \gamma_2 [\Psi^{+}_{F\sigma}(x=0) c_{-\sigma}(x=0) + \text{H.c.}]. \tag{3.24}$$

The time-scale for the spin flips will be given by the relevant energy scale $h_{\text{ex}}$ which is large compared to all other energy scales, so that it is sensible to also assume this process to be instantaneous. Adding $H_{T2}$ to the system's Hamiltonian in Eq. (3.20) we need to introduce a second contribution to Eq. (2.6) to access the CGF

$$\chi_{\text{SFa}}(\lambda) = \left\langle T_C \exp[-i \int_C dt (T_1^\lambda + T_2^\lambda)] \right\rangle, \tag{3.25}$$

where $T_2^\lambda$ denotes $H_{T2}$ with the additional substitution $\Psi_{F\sigma}(x=0) \rightarrow \Psi_{F\sigma}(x=0) e^{-i\lambda(t)/2}$ as done for $H_{T1}$ in Eq. (3.4).
The CGF for a SC-FM-QPC including spin-active scattering is quite complicated in full detail and we only want to give here its simplified form using again the approximation in Eqs. (3.14) and

(3.15). We arrive at

$$
\begin{aligned}
\ln \chi_{\mathrm{SF,f}}(\lambda) &= 2\tau \int \frac{d\omega}{2\pi} \Big[ \ln \Big( \{1 + T_{\mathrm{e}\uparrow,\mathrm{f}}[n_{F+}(1-n_S)(e^{i\lambda}-1) + n_S(1-n_{F+})(e^{-i\lambda}-1)]\} \\
& \{1 + T_{\mathrm{e}\downarrow,\mathrm{f}}[n_{F+}(1-n_S)(e^{i\lambda}-1) + n_S(1-n_{F+})(e^{-i\lambda}-1)]\} \\
& -T_{\mathrm{d,f}}[n_{F+}(1-n_S)(e^{i\lambda}-1) + n_S(1-n_{F+})(e^{-i\lambda}-1)]^2 \\
& -T_{\mathrm{s,f}}[n_{F+}(1-n_S)(e^{i\lambda}-1) + n_S(1-n_{F+})(e^{-i\lambda}-1)] \Big) \theta\left(\frac{|\omega|-\Delta}{\Delta}\right) \\
& +1/2 \ln \Big( \{1 + T_{\mathrm{A,f}}[n_{F+}(1-n_{F-})(e^{2i\lambda}-1) + n_{F-}(1-n_{F+})(e^{-2i\lambda}-1)]\}^2 \\
& -T_{\mathrm{A2,f}}[n_{F+}(1-n_{F-})(e^{2i\lambda}-1) + n_{F-}(1-n_{F+})(e^{-2i\lambda}-1)] \\
& +T_{\mathrm{AT,f}}[n_{F+}(1-n_{F-})(e^{2i\lambda}-1) + n_{F-}(1-n_{F+})(e^{-2i\lambda}-1)] \Big) \theta\left(\frac{\Delta-|\omega|}{\Delta}\right) \Big] \\
& .
\end{aligned}
\tag{3.26}
$$

We define the effective transmission coefficients to be

$$
\begin{aligned}
T_{\mathrm{e}\sigma,\mathrm{f}} &= \frac{4(\Gamma_{11\sigma}+\Gamma_{12\sigma})}{(1+\Gamma_{11\uparrow}+\Gamma_{12\uparrow})(1+\Gamma_{11\downarrow}+\Gamma_{12\downarrow})-\Gamma_{13\uparrow}\Gamma_{13\downarrow}}, \ \sigma=\uparrow,\downarrow \\
T_{\mathrm{d,f}} &= \frac{16\Gamma_{13\uparrow}\Gamma_{13\downarrow}}{[(1+\Gamma_{11\uparrow}+\Gamma_{12\uparrow})(1+\Gamma_{11\downarrow}+\Gamma_{12\downarrow})-\Gamma_{13\uparrow}\Gamma_{13\downarrow}]^2}, \\
T_{\mathrm{s,f}} &= \frac{4[(\Gamma_{11\uparrow}-\Gamma_{11\downarrow}+\Gamma_{12\uparrow}-\Gamma_{12\downarrow})^2+\Gamma_{13\uparrow}\Gamma_{13\downarrow}]}{[(1+\Gamma_{11\uparrow}+\Gamma_{12\uparrow})(1+\Gamma_{11\downarrow}+\Gamma_{12\downarrow})-\Gamma_{13\uparrow}\Gamma_{13\downarrow}]^2}, \\
T_{\mathrm{A,f}} &= \frac{4[(\Gamma_{21\downarrow}+\Gamma_{22\downarrow})(\Gamma_{21\uparrow}+\Gamma_{22\uparrow})-\Gamma_{23\uparrow}\Gamma_{23\downarrow}]}{W}, \\
T_{\mathrm{A2,f}} &= \frac{4(\Gamma_{23\uparrow}+\Gamma_{23\downarrow})^2(\Gamma_{21\uparrow}+\Gamma_{22\uparrow})(\Gamma_{21\downarrow}+\Gamma_{22\downarrow})}{W^2}, \\
T_{\mathrm{AT,f}} &= \frac{4(\Gamma_{23\uparrow}+\Gamma_{23\downarrow})^2[1+\Gamma_{23\uparrow}\Gamma_{23\downarrow}]^2}{W^2}, \\
W &= 1+(\Gamma_{21\uparrow}+\Gamma_{22\uparrow})(\Gamma_{21\downarrow}+\Gamma_{22\downarrow})[2+(\Gamma_{21\uparrow}+\Gamma_{22\uparrow})(\Gamma_{21\downarrow}+\Gamma_{22\downarrow})] \\
& -2(\Gamma_{21\uparrow}+\Gamma_{22\uparrow})(\Gamma_{21\downarrow}+\Gamma_{22\downarrow})\Gamma_{23\uparrow}\Gamma_{23\downarrow}+(1+\Gamma_{23\uparrow})^2(1+\Gamma_{23\downarrow})^2.
\end{aligned}
$$

In these definitions we used the abbreviations

$$
\begin{aligned}
\Gamma_{11\sigma} &= \frac{\Gamma(1+\sigma P)|\omega|}{\sqrt{\omega^2-\Delta^2}}, \ \Gamma_{12\sigma} = \frac{\Gamma_f(1+\sigma P)|\omega|}{\sqrt{\omega^2-\Delta^2}}, \\
\Gamma_{13\sigma} &= \frac{2(1+\sigma P)(\Gamma\Gamma_f)^{1/2}|\omega|}{\sqrt{\omega^2-\Delta^2}}, \ \Gamma_{21\sigma} = \frac{\Gamma(1+\sigma P)\Delta}{\sqrt{\Delta^2-\omega^2}}, \\
\Gamma_{22\sigma} &= \frac{\Gamma_f(1+\sigma P)\Delta}{\sqrt{\Delta^2-\omega^2}}, \ \Gamma_{23\sigma} = \frac{2(1+\sigma P)(\Gamma\Gamma_f)^{1/2}\Delta}{\sqrt{\Delta^2-\omega^2}},
\end{aligned}
$$

with $\Gamma_f = \rho_{0F}\rho_{0S}\gamma_2^2\pi^2/2$.
We observe a more complicated structure of the CGF compared to Eq. (3.22) since we do not only need to introduce a spin-dependent DOS but also two contact transparencies $\Gamma$, $\Gamma_f$ that refer to the normal and spin-flipped transparency, respectively. Above the gap we observe single-electron transmission for the different spins described by $T_{\mathrm{e}\sigma,\mathrm{f}}$. Additionally, spin-flip transmission of single electrons must contribute giving rise to the transmission coefficients $T_{\mathrm{d,f}}$ and $T_{\mathrm{s,f}}$, see Fig 3.2 (c) and (d). In the numerator of $T_{\mathrm{s,f}}$ there are two contributions since one additionally has to keep track of the asymmetric DOS for the different spins. Below the gap we find two types of AR: $T_{\mathrm{A,f}}$

and $T_{\mathrm{A2,f}}$ describe the normal, spin-symmetric AR and $T_{\mathrm{AT,f}}$ describes spin-flipped AR (SAR), see Fig. 3.2 (a) and (b). The latter involves a spin-flip process during the AR.
We have obtained the FCS of all charge transfer processes that have also been identified in the quasi-classical Green's function calculation (Grein et al. [ 2010]). The only difference is the description of spin-active scattering. Grein et al. use the spin-mixing angle $\theta_s$ as the phenomenological parameter, whereas we use a second tunneling transparency to account for spin-flips. Both descriptions are related since both give rise to Andreev bound states characterised by $T_{\mathrm{A,f}}(\epsilon_\pm) = 1$ from which one can calculate $\theta_s$ via

$$\epsilon_\pm = \pm\Delta\cos(\theta_s/2). \tag{3.27}$$

Apart from the quasi-classical Green's function formalism (Grein et al. [ 2010]) and the approach presented here a third theoretical treatment has been frequently used for the analysis of SC-FM-QPCs: the extended BTK model (Martín-Rodero et al. [ 2001], Pérez-Willard et al. [ 2004]). This model has been very successful for certain experiments but fails for others.

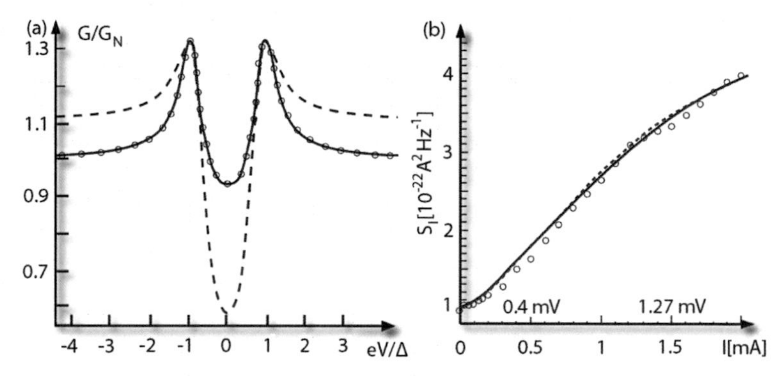


**Figure 3.3.: (a)**: Experimental data for the differential conductance as a function of $V$ of a SC-FM-QPC taken from Pérez-Willard et al. [ 2004] (black circles). The data has been normalised with respect to the normal state conductance $G_N$. The blue solid curve is the prediction using Eq. (3.26) for $T = 0.1\Delta$, $\Gamma = 0.095$, $\Gamma_f = 0.065$, $P = 0.38$ and a broadening of the BCS DOS described by $\Gamma_D = 0.09\Delta$. One observes the characteristic double peak structure at the SC gap. We also compare this result to the best possible fit without spin-active scattering using $T = 0.13\tilde{\Delta}$, $\Gamma = 0.13$, $\Gamma_f = 0$, $P = 0.38$, $\Gamma_D = 0.09\tilde{\Delta}_0$ and $\tilde{\Delta} = 0.75\Delta$ (dashed red curve). Without spin-active scattering we have to choose a gap $\tilde{\Delta}$ different from the one obtained in the experiment in order to obtain the conductance peaks at the correct position.
**(b)**: Theoretical prediction for the noise (blue dashed curve) of the SC-FM-QPC for $T = 0.1\Delta$, $\Gamma = 0.14$, $\Gamma_D = 0.2\Delta$ and $P = 0.42$ using Eq. (B.1). The theoretical prediction for the case of a SC-normal metal-QPC using the CGF in Eq. (3.11) for $T = 0.1\Delta$, $\Gamma = 0.14$ and $\Gamma_D = 0.005\Delta$ is shown in red. In both cases one observes the transition from two-electron tunneling to single-electron tunneling for voltages above the SC gap that is located at $V = 1.35$ mV. Both predictions agree very well with the experimental data for a SC-normal conductor junction taken from Jehl et al. [ 2000] shown as black circles. The universal shape of the noise as a function of current is lost if the interface-transparency is high enough to see effects from the energy-dependent DOS of the SC.

Here we show that our model (as the quasi-classical Green's function formalism) reproduces the experimental data from (Pérez-Willard et al. [ 2004]) for a finite spin-flip amplitude.

The differential conductance is the derivative of the current which in turn is given by the first derivative of the CGF with respect to the counting field $I_{\mathrm{SF,f}} = -i/\tau\partial \ln \chi_{\mathrm{SF,f}}(\lambda)/\partial\lambda|_{\lambda=0}$. We compare the differential conductance $dI_{\mathrm{SF,f}}/dV$ to the experimental data for Al/Co contacts. The result is shown in Fig. 3.3 (a).

Perfect agreement is obtained for a reasonable Dynes parameter (meaning a much smaller value than the gap), which again signifies the importance of spin-active scattering for strongly polarized FMs as Co. The result incorporating spin-active scattering may also be compared to the best possible fit without spin-active scattering (see Fig. 3.3 (a)). We see that a consistent explanation of the experimental data heavily relies on the inclusion of a complete description of the interface region. The possibility of SAR opens a new transport channel below the gap that leads to an increased conductance for $V < \Delta$. Indeed the contribution by SAR is sizeable since the ratio of SAR vs. AR is determined by the ratio of $\Gamma_f$ and $\Gamma$ being of the order of 0.7. This is in accordance with other studies of different point contacts (Grein et al. [ 2010], Hübler et al. [ 2012]).

The FCS allows for the calculation of the noise power given by the second derivative of the CGF $S_{\mathrm{SF,f}} = -1/\tau\partial^2 \ln \chi_{\mathrm{SF,f}}(\lambda)/\partial\lambda^2|_{\lambda=0}$. We only analyse the noise as a function of the current in the regime of small interface transparency. In this case the energy dependence of the transmission coefficients plays only a minor role. The noise should show a smooth transition from AR noise (corresponding to a Fano factor of 2) below the gap to single-electron noise above the gap (corresponding to a Fano factor of 1), since these represent the only dominant charge transfer processes in these regimes. Consequently the form of $S(I)$ should be identical to the one obtained for SC-normal metal-QPCs since the charge transfer processes in both systems are the same. The universal shape remains even in the presence of SAR and spin-flip transmission since these do not cause a change of the Fano factor in the respective energy regimes. SARs are two-electron processes as ARs and spin-flip transmission is a single-electron process as simple electron transmission.

This discussion can be extended to diffusive charge transfer only under certain assumptions since the analysis of the FCS for diffusive structures in Belzig and Nazarov [ 2001a] revealed that one expects an enhancement of the noise at voltages of the order of and for temperatures below $E_{\mathrm{Th}}$, the Thouless energy. $E_{\mathrm{Th}} = D/L^2$ can be calculated from the diffusion coefficient $D$ of the electrons and the size $L$ of the sample. Additionally one has to mind that diffusive samples are usually in the limit of a large number of (rather opaque) transmission channels and the disorder in the normal metal could play a role.

Indeed, noise measurements have been performed on diffusive SC-normal samples (Jehl et al. [ 2000]) so that a direct test of the above discussed universal shape of $S(I)$ for small interface transparency becomes possible. For the experiment under consideration (Jehl et al. [ 2000]) we find that$E_{\mathrm{Th}} \approx 0.05$ meV which is small compared to the voltages applied. Therefore, we only probe the noise in the incoherent semiclassical limit and according to (Belzig and Nazarov [ 2001a]), coherence between the charge carriers should play no role.

To visualize the general behavior conjectured above, we take the same transmission coefficient and temperature in the case of a SC-normal metal-QPC and a SC-FM-QPC and use a typical Dynes parameter and polarisation for a FM to make a prediction for the noise (Fig. 3.3 (b)). We compare both results to the experimental data in the case of a diffusive contact. The noise spectrum shows the same behavior in the diffusive limit and the ballistic limit independent of the nature of the second lead being a normal metal or a FM, however, only in the limit of small transmission. The noise spectrum for the FCS with spin-active scattering also shows the same behavior.

## 3.4. Conclusions

In conclusion, we have demonstrated how to access the CGF using the Hamiltonian approach for SC hybrid structures. For the case of a SC-normal metal tunnel contact we reproduce previous results obtained using different methods and have discussed possible approximations of the CGF. Afterwards we have turned to the study of SC-FM-QPCs. We observe that the assumption of simple spin-conserving tunneling is not adequate in this case and have extended our model in order to incorporate effects from spin-active scattering. We have shown how to calculate the CGF and demonstrate the perfect reproduction of previous results and experimental data. We have also discussed the noise properties of SC hybrid sturctures in the limit of small transparency and compare our results to experimental data for diffusive junctions.

# Chapter 4

## Majorana fermions

In Chapter 3 in Section 3.3 we discussed how $p$-wave correlations can be generated in a FM using spin-active scattering and the proximity coupling to a SC. In Chapter 2 in Section 2.6 we discussed the possibility of realizing Majorana fermion bound states in a nanowire proximity coupled to a $p$-wave SC and also discussed how to mimick the behavior of this system using an $s$-wave SC in combination with spin-orbit coupling and a magnetic field along the wire direction. In this Chapter we want to demonstrate that one may also use the aforementioned $p$-wave correlations in a FM in a possibly even simpler setup. In the second part of this Chapter we will discuss how to detect the Majorana fermion states unambiguously using the CGF of charge transfer through one Majorana fermion bound state. Again, we will use the Hamiltonian formalism introduced before and follow the analysis in Soller et al. [ 2012b]. Finally, we will show how Majorana fermions can also be realized in exciton condensates (ExCs) following Soller and Komnik [ 2013].

### 4.1. Majorana fermions in ferromagnetic wires

We consider the setup depicted in Fig. 4.1 : a ferromagnetic wire (FM wire) is deposited on a SC and further coupled to a STM tip. Since the contact between the SC and the FM wire is much better than the one between the FM wire and the STM tip we consider a voltage drop only between the latter two parts of the system. We approach the coupling of the SC to the FM wire in two steps: first, we describe a lateral tunnel contact in order to extract the general gap structure induced in the wire due to the presence of the SC. In a second step we use this description in order to write down a low-energy effective Hamiltonian describing the wire. The starting point for the first part is the tunnel contact between the SC and the FM-wire,

$$\tilde{H}_{\text{wire}} = H_F + H_S + H_{\text{T,1D}}. \tag{4.1}$$

Concerning the tunneling between the SC and the FM wire we have to mind that the wire has a

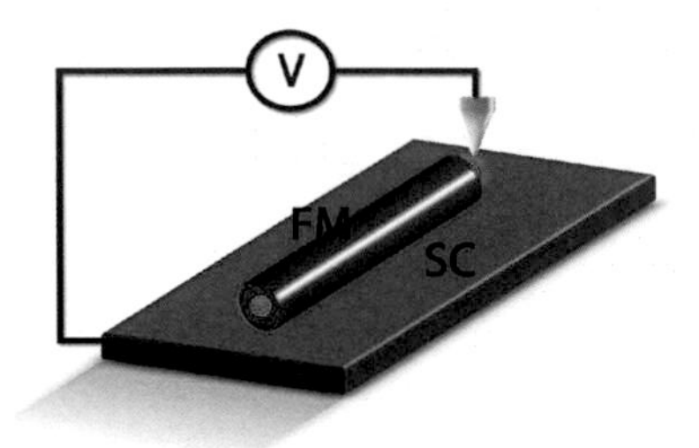


**Figure 4.1.:** Setup for the generation of Majorana fermions: a SC is tunnel coupled to a FM wire. Due to the induced triplet superconductivity the wire hosts a Majorana bound state (MBS) at the end whose transport characteristics are investigated when coupled to an STM tip at the wire end. The wire lies completely on the substrate. The coupling between the SC and the FM wire is much better than the one between the MBS and the STM tip so that a voltage ($V$) drop is assumed only to occur between these two parts of the system.

lateral extension and we cannot use the simple tunneling Hamiltonian as in Eq. (2.3). Instead we have to keep track of momentum conservation along the wire direction. Additionally, we found in Section 3.3 that we have to include the spin-activity of the interface in SC-FM hybrids. Again, we do so by introducing a second spin-active tunneling term in $H_{\mathrm{T,1D}}$

$$H_{\mathrm{T,1D}} = \sum_{\mathbf{k},\sigma} \gamma_{1,\mathrm{1D}}[\Psi^{+}_{\mathbf{k}\| F\sigma} c_{\mathbf{k}\sigma} + \mathrm{H.c.}] + \sum_{\mathbf{k},\sigma} \gamma_{2,\mathrm{1D}}[\Psi^{+}_{\mathbf{k}\| F\sigma} c_{\mathbf{k}-\sigma} + \mathrm{H.c.}], \tag{4.2}$$

where $\mathbf{k}\|$ refers to the momentum along the wire and $\mathbf{k}\perp$ refers to the momenta in perpendicular direction. $\mathbf{k}$ is the full momentum. As in (Chung et al. [ 2011]) the model can be solved exactly as it is quadratic in fermion fields. Apart from the induced $s$-wave pairing in the FM wire describing induced correlations of the type $\langle \Psi^{+}_{\uparrow} \Psi^{+}_{\downarrow} \rangle$ we find additional $p$-wave correlations $\langle \Psi^{+}_{\uparrow,\downarrow} \Psi^{+}_{\uparrow,\downarrow} \rangle$ that can be traced back to the spin-activity of the interface as in Section 3.3 . However, due to the exchange splitting of the FM wire $s$-wave correlations and $p$-wave correlations of the type $1/\sqrt{2}(|\uparrow\downarrow\rangle + |\downarrow\uparrow\rangle)$ will decay on a length scale $\hbar/(p_{F\uparrow} - p_{F\downarrow})$ with $p_{F\sigma}$ referring to the Fermi momentum of the electrons with spin $\sigma$. Consequently, only the equal spin correlations will remain (Grein et al. [ 2010]). The p-wave correlations $\langle \Psi^{+}_{\uparrow,\downarrow} \Psi^{+}_{\uparrow,\downarrow} \rangle$ will penetrate into the FM on a length scale $\hbar v_{F,\uparrow,\downarrow}/\Delta$. Therefore, we assume the wire thickness $d$ to be $\hbar/(p_{F\uparrow} - p_{F\downarrow}) \ll d < \hbar v_{F,\uparrow,\downarrow}/\Delta$.
A solution by diagonalization of the Hamiltonian is possible via considering a 1D SC coupled to a FM wire and averaging over the additional momenta $\mathbf{k}\perp$ in the SC afterwards. In this case our model Hamiltonian in the 1D case can be written as

$$H_{1D} = \frac{1}{2} \sum_{\mathbf{k}\|} \Psi^{+}_{\mathbf{k}\|} H^{\mathrm{BdG}}_{\mathbf{k}\|} \Psi_{\mathbf{k}\|}, \text{ where } H^{\mathrm{BdG}}_{\mathbf{k}\|} = \begin{bmatrix} h^{\mathrm{FM}} & T_{\mathbf{k}\|} \\ T^{+}_{\mathbf{k}\|} & \Lambda_{\mathrm{BdG}} \end{bmatrix}, \tag{4.3}$$

using $\Psi_{\mathbf{k}\|} = (\Psi_{\mathbf{k}\| F\uparrow}, \Psi_{\mathbf{k}\| F\downarrow}, \Psi^{+}_{-\mathbf{k}\| F\uparrow}, \Psi^{+}_{-\mathbf{k}\| F\downarrow}, B_{\mathbf{k}\|,\uparrow}, B_{\mathbf{k}\|,\downarrow}, B^{+}_{-\mathbf{k}\|,\uparrow}, B^{+}_{-\mathbf{k}\|,\downarrow})$ and $h^{\mathrm{FM}}$ referring to Eq. (3.21) written in terms of the introduced $\Psi_{\mathbf{k}\|}$ fields. $T_{\mathbf{k}\|}$ and $\Lambda_{\mathrm{BdG}}$ refer to Eqs. (4.2) and (2.7) written in terms of $\Psi_{\mathbf{k}\|}$ fields where $B_{\mathbf{k}\|,\sigma}$ operators refer to Bogoliubov quasiparticle annihilation operators introduced in Appendix A Eq. ( A.16).
The Hermitian matrix $H^{\mathrm{BdG}}_{\mathbf{k}\|}$ can now be diagonalized $\left(U^{+}_{\mathbf{k}\|} H^{\mathrm{BdG}}_{\mathbf{k}\|} U_{\mathbf{k}\|}\right)$ with the unitary matrix $U_{\mathbf{k}\|}$. This procedure (Chung et al. [ 2011]) allows to express the pairing amplitude in terms of elements

of $U_{\mathbf{k}||}$

$$\langle\Psi_{-\mathbf{k}||F\uparrow}\Psi_{\mathbf{k}||F\uparrow}\rangle = (U_{\mathbf{k}||})^*_{33}(U_{\mathbf{k}||})_{13} + (U_{\mathbf{k}||})^*_{34}(U_{\mathbf{k}||})_{14} + (U_{\mathbf{k}||})^*_{37}(U_{\mathbf{k}||})_{17} + (U_{\mathbf{k}||})^*_{38}(U_{\mathbf{k}||})_{18}.$$

Averaging this expression over $\mathbf{k}\perp$ leads to a momentum dependent pairing gap $\Delta_{\uparrow\uparrow}(k)$. However, the general solution is quite complicated. Since, for the low-energy behavior relevant for the presence of a Majorana fermion, only low momenta are relevant we can consider the case $k = 0$ and obtain the following result for $h_{\text{ex}} = 0$ and $\gamma_{2,1\text{D}} \neq 0$

$$\frac{\Delta_{\uparrow\uparrow}}{\Delta} = \frac{2 + \gamma_{1,1\text{D}}^2 + \gamma_{2,1\text{D}}^2 - 2\sqrt{1 + \gamma_{1,1\text{D}}^2 + \gamma_{2,1\text{D}}^2}}{\gamma_{2,1\text{D}}^2}. \tag{4.4}$$

In this case for $h_{\text{ex}} = 0$ of course $\Delta_{\uparrow\uparrow} = \Delta_{\downarrow\downarrow}$. For finite $h_{\text{ex}}$ the two proximity induced gaps are different. Following (Gangadharaiah et al. [2011], Oreg et al. [2010], Volkov et al. [1995]) the effective low-energy Hamiltonian describing the proximity effect should be a standard BCS Hamiltonian, however, with two proximity induced $p$-wave gaps $\Delta_{\uparrow\uparrow,\downarrow\downarrow}/\Delta$. This consideration generalizes the result obtained in (Chung et al. [2011], Duckheim and Brouwer [2011]): in the case of a ferromagnetic halfmetal no spin-$\downarrow$ exists and only one $p$-wave gap survives, which leads to the emergence of Majorana bound states (MBSs). However, for a finite exchange field in the FM always both spin species will be present.
In this case we might still obtain a MBS but via a mechanism more akin to the one in InAs/InSb nanowires discussed in Section 2.6, where one uses spin-orbit coupling in combination with an applied magnetic field. In our proposal we can use spin-active scattering and ferromagnetism in the same way. Following our description above, the simplest Hamiltonian describing the FM wire in the presence of the SC reads

$$\begin{aligned} H_{\text{wire}} &= \int dx \Psi^+(x)\mathcal{H}\Psi(x),\ \Psi^+ = (\Psi_\uparrow^+, \Psi_\downarrow^+, \Psi_\downarrow, \Psi_\uparrow), \\ &\text{where } \mathcal{H} = \left[\frac{p^2}{2m} - \mu_{\text{wire}}\right]\tau_z - h_{\text{ex}}\sigma_z + \begin{pmatrix} 0 & \Delta_{\uparrow\uparrow} \\ \Delta_{\downarrow\downarrow} & 0 \end{pmatrix}\sigma_x. \end{aligned}$$

$\Psi_{\uparrow,\downarrow}(x)$ annihilates spin-$\uparrow$ ($\downarrow$) electrons at position $x$. The Pauli matrices $\sigma$ and $\tau$ operate in spin- and particle-hole space, respectively. $\mu_{\text{wire}}$ is the chemical potential, that we choose to be zero.
The spectrum can be revealed as in (Gangadharaiah et al. [2011], Oreg et al. [2010]) by squaring $\mathcal{H}$ twice, which yields

$$E_{\sigma,\pm} = \sigma h_{\text{ex}} \pm \sqrt{\Delta_{\sigma\sigma}^2 + \xi_p^2}, \tag{4.5}$$

where $\xi_p = p^2/(2m)$. As in the aforementioned works the gap $E_0$ near $p = 0$ is the key to the emergence of MBSs. We find

$$E_0 = h_{ex} - \Delta_{\uparrow\uparrow}. \tag{4.6}$$

In the proposal using spin-orbit coupled wires two regimes corresponding to the topologically trivial and nontrivial case exist depending on the relative strength of the magnetic field compared to the proximity induced gap. In our case $h_{\text{ex}} \gg \Delta_{\uparrow\uparrow,\downarrow\downarrow}$ for typical FMs. Therefore, we always obtain an exchange field-dominated (or strong interaction induced) gap and hence the wire will be in its topological phase. The end of the wire can now be characterised by a sharp drop of the chemical potential, which closes the gap. Since this transition corresponds to a transition out of the topological phase MBSs will be localized at the wire ends.

## 4.2. Detection of Majorana fermions

In Section 4.1 of this Chapter we have described how to create a pair of Majorana fermions in a FM wire. In this Section we want to consider a possible detection scheme for the Majorana fermion. The Majorana fermion creates a zero-energy bound state in the FM wire (or equivalently the strongly spin-orbit coupled nanowire). The bound state can be observed in the differential conductance as a peak corresponding to perfect conductance at zero bias. However, this observation alone is not enough since it may also be caused by a Kondo resonance, see Section 2.4. Therefore, it has to be accompanied e.g. by observing the Josephson current to make sure that the peak is indeed caused by a Majorana fermion (Oreg et al. [2010]). In this Section we will show that a similar conclusion can be obtained by measuring higher cumulants of charge transfer.
Using one of the above described proposals for the generation of Majorana fermions the effective low-energy behavior of the system is that of a $p$-wave SC. Therefore the FM coupling to the STM tip may be treated as an effective $p$-wave SC with the Hamiltonian (Gangadharaiah et al. [2011])

$$H_{\text{eff}} = \sum_k \epsilon_k \Psi^+_{kP\uparrow}\Psi_{kP\uparrow} + \sum_k (\Delta_p \Psi^+_{kP\uparrow}\Psi^+_{-kP\uparrow} + \Delta_p^* \Psi_{-kP\uparrow}\Psi_{kP\uparrow}), \tag{4.7}$$

where $\Delta_p$ refers to the effective $p$-wave gap. Compared to the $s$-wave SC in Eq. (2.7) we now break time-reversal symmetry and spin-rotation symmetry. Electron-hole symmetry at the Fermi level remains and we have achieved the typical Majorana situation (Mudry et al. [2000]).
We assume from now on to have a MBS at the wire end and describe its detection via coupling to the STM tip also indicated in Fig. 4.1. The tip is described as a normal metal with a flat band DOS $\rho_{0T}$ and chemical potential $\mu_T$ using electron field operators $\Psi_{kT\sigma}$. The FM wire in Eq. (4.7) is held at chemical potential $\mu_{\text{wire}} = 0$. Tunneling between the FM wire in Eq. (4.7) and the STM tip is given by the usual tunneling Hamiltonian as in Eq. (2.3)

$$H_{\text{T,STM}} = \gamma_T[e^{i\tilde{\phi}/2}\Psi^+_{P\uparrow}(x=0)\Psi_{k,T,\uparrow}(x=0) + \text{H.c.}], \tag{4.8}$$

which, however, includes a possible phase shift $\tilde{\phi} = \pm\phi$ for electrons/holes, to account for the different topological phases of the system. We want to describe the situation where the effective $p$-wave SC hosts a MBS, which represents the topologically non-trivial phase of this system. A topological phase transition is characterised by a change in the topological quantum number $Q_{\text{top}}$. In our case the topological quantum number is the scattering phase shift during AR from a Majorana fermion (Akhmerov et al. [2011]). Either $\phi = 0$ or $\phi = \pi$ in the topologically trivial/non-trivial case, respectively. Consequently, in the topologically non-trivial phase, the two Andreev bound states at $\pm\Delta_p$ that are present in the topologically trivial phase (see Eq. (3.27)) merge (also in non-equilibrium) into one bound state at

$$\epsilon_{\text{MBS}} = \pm\Delta_p\cos(\pi/2) = 0.$$

From here we use again the substitution in Eq. (2.5) and calculate the CGF using the Hamiltonian approach as in Section 3.1. The CGF in the limit of low tunnel coupling $\gamma_T$ will be the sum of two contributions for the different energy regimes with respect to the $p$-wave SC gap as in Section 3.1

$$\ln\chi_M(\lambda) = \ln\chi_{\text{e,M}}(\lambda) + \ln\chi_{\text{A,M}}(\lambda). \tag{4.9}$$

The zero-energy Majorana bound state affects the properties of AR in our system. However, for the FCS of charge transfer above the gap we may neglect AR and branch-crossing due to the small tunnel coupling of the STM tip and arrive at Eq. (3.18) rephrased in terms of the variables introduced here

$$\ln \chi_{\rm e,M}(\lambda) = \tau \int \frac{d\omega}{2\pi} \ln\{1 + T_{\rm e,M}(\omega)[n_{T+}(1-n_F)(e^{i\lambda}-1) + n_F(1-n_{T+})(e^{-i\lambda}-1)]\} \times\theta\left(\frac{\omega-\Delta_p}{\Delta_p}\right),$$

where $n_{T+}$ and $n_F$ refer to the Fermi distributions of the tip and the effective $p$-wave SC respectively. $T_{\rm e,M}(\omega) = T_{\rm e,P}|\omega|/\sqrt{\omega^2-\Delta_p^2}$, $T_{\rm e,P} = 4\Gamma(1+P)$, where $P$ refers to the polarisation of the FM wire. Finally we get to the contribution below the gap that allows to observe the topologically non-trivial phase

$$\ln \chi_{\rm A,M}(\lambda) = \tau \int \frac{d\omega}{2\pi} \ln\{1 + T_{\rm A,M}(\omega)[(e^{2i\lambda}-1)n_{T+}(1-n_{T-}) + (e^{-2i\lambda}-1)n_{T-}(1-n_{T+})]\} \times\theta\left(\frac{\Delta_p-\omega}{\Delta_p}\right), \quad (4.10)$$

$$T_{\rm A,M}(\omega) = \frac{T_{e,P}^2}{1+R_{e,P}^2 - 2R_{e,P}\cos(2\arccos(\omega/\Delta_p)+\pi)}, \quad (4.11)$$

where $R_{\rm e,P} = 1 - T_{\rm e,P}$ and $n_{T-} = 1 - n_{T+}(-\omega)$ refers to the Fermi distribution of the holes. The additional phase factor $\pi$ in Eq. (4.11) is the difference to the topologically trivial case where only the Andreev reflection phase shift $\arccos(\omega/\Delta_p)$ would be present (Hübler et al. [ 2012]).
Next, we study the consequences of the additional phase factor. We obtain a perfectly transmitting channel at $\omega = 0$ which corresponds to a zero-bias conductance of $2e^2/h$ in SI units at zero temperature that can be calculated from the first derivative of the CGF and agrees with previous studies (Flensberg [ 2010]).

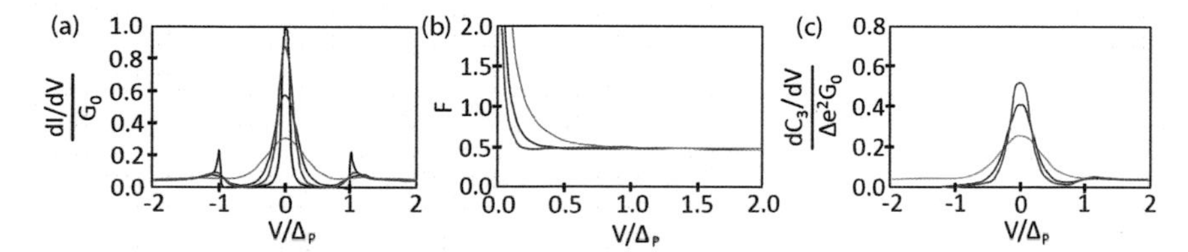


**Figure 4.2.:** Results derived from the CGF in Eq. (4.9) for the conductance, Fano factor and differential third cumulant for $\Gamma(1+P) = 0.01$ and $T = 0.06\Delta$ (red), $T = 0.1\Delta$ (blue) and $T = 0.2\Delta$ (green): in **(a)** we show the result for the conductance. The black curve is the result for $T = 0.01\Delta$, where we observe that the conductance reaches (almost) $G_0$. Furthermore we see the proximity induced SC correlations at $V = \pm\Delta_p$. The Fano factor in **(b)** at $V = 0$ shows only thermal noise and then quickly goes to 0.5 describing charge transport of a single spin species. Part **(c)** shows the differential third cumulant that follows the form of the conductance in **(a)**.

In Fig. 4.2 we plotted the results for the conductance, Fano factor$F = S/(2 \cdot I)$ and the differential third cumulant that one obtains by taking the first, second and third derivative with respect to $\lambda$ of Eq. (4.9). The figure shows the results for three different temperatures. We see that while observing the SC correlations in the conductance (Fig. 4.2 (a)) really requires to go to low temperatures with respect to the proximity induced gap, the features caused by the MBS remain stable upon varying temperature. Consequently, everything that is crucial, is to induce a sizeable $p$-wave gap in the FM

wire.
Concerning noise we see that, the MBS opens a perfectly transmitting channel and consequently we only observe thermal noise at $V = 0$. For finite voltage the Fano factor quickly approaches 0.5 referring to the charge transfer of 'half a fermion'. The differential third cumulant follows a Levitov-Reznikov-like behavior (Levitov and Reznikov [ 2004]) since it follows the form of the conductance $dC_3/dV \propto dI/dV$.
These three signatures: a peak in the conductance, the form of the Fano factor and the behavior of the third cumulant give a very clear prediction to be searched for in future experiments. States at the interface that are not bound states could show a similar pattern in the conductance but not in the Fano factor and the third cumulant. The only states that could show a similar behavior are Andreev bound states or Kondo resonances. The former have so far have only been observed in SC-FM heterostructures. Additionally Andreev bound states would not be localised at the end of the wire. Therefore moving the tip away from the end of the wire must lead to the disappearance of the peak in the conductance and therefore allows for an unambiguous identification of the MBS. Also, a Kondo resonance could give a similar feature in the conductance but the Fano factor would be different (Gogolin and Komnik [ 2006a], Soller and Komnik [ 2011a]), see also Chapter 6 .

## 4.3. Majorana fermions in exciton condensates

Finally, we would like to illustrate still a fourth possibility for the generation of Majorana fermions, which does not rely on the proximity effect of a BCS SC as the ones we have illustrated in Sections 2.6 and 4.1 . In Section 2.5 we introduced yet another BCS condensate: an exciton condensate (ExC). Also ExCs can host Majorana fermions as has been shown in Soller and Komnik [ 2013].
We use the setup as depicted in Fig. 4.3 : the bilayer ExC is assumed to be spin-polarized. Such a situation may be created e.g. by applying a sufficiently large magnetic field (Lozovik et al. [ 2002]). Additionally the ExC is proximity coupled to two $p$-wave SCs.

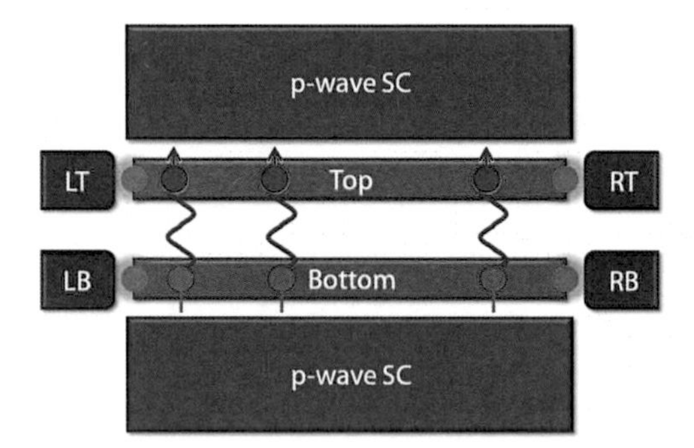


**Figure 4.3.:** Setup considered for the generation of Majorana fermions in ExCs. A bilayer ExC is spin-polarized by a magnetic field (indicated by the arrows for the particles in the top and bottom layer). Additionally the ExC is proximity coupled to two $p$-wave SCs. The emerging Majorana fermions are indicated as yellow spots in the top and bottom layer.

We want to show that this system has the same low energy behavior as the one discussed for InAs/InSb nanowires in Section 2.6 . This time we use the spinless nature of the ExC in combination with the $p$-wave SC. The simplest Hamiltonian describing the ExC in the presence of the $p$-wave SC reads combining Eq. (2.10) and the proximity induced $p$-wave coupling $\Delta_p$ (Volkov et al. [1995])

$$H_{\text{ExC,M}} = \int dx \Psi^{+}_{\text{ExC}}(x)\mathcal{H}_{\text{ExC,M}}\Psi_{\text{ExC}}(x), \; \Psi^{+}_{\text{ExC}} = (\Psi^{+}_{T}, \Psi^{+}_{B}, \Psi_{B}, \Psi_{T}), \tag{4.12}$$

$$\text{where } \mathcal{H}_{\text{ExC,M}} = \left[\frac{p^2}{2m}\right]\tau_z + \Delta\tau_x + \Delta_p\sigma_x\tau_x. \tag{4.13}$$

The Pauli matrices $\tau$ and $\sigma$ operate in particle-hole and top-bottom space, respectively.
The spectrum can again be revealed as in (Gangadharaiah et al. [2011], Oreg et al. [2010]) by squaring $\mathcal{H}_{\text{ExC,M}}$ twice, which yields

$$E^2_{\text{ExC},\pm} = \Delta_p^2 + \Delta^2 + \xi_p^2 \pm (2\Delta\Delta_p), \tag{4.14}$$

where $\xi_p = p^2/(2m)$. As before the gap $E_{\text{ExC},0}$ near $p = 0$ is the key to the emergence of MBSs. We find

$$E_0 = |\Delta_p - \Delta|. \tag{4.15}$$

Consequently we observe an exchange field-dominated (or strong interaction-induced) gap if $\Delta_p > \Delta$ and hence the ExC in this case will host Majorana fermions. The end of the ExC can again be characterised by a sharp drop of the chemical potential, which closes the gap. Since this transition corresponds to a transition out of the topological phase MBSs will be localized at the ExC ends. However, this was to expected, since we have basically created a system consisting of two times the Kitaev model discussed in Section 2.6. The only difference to the former proposal is that now we will have four Majorana fermions instead of two, however, in a physically very different system. The situation is slightly different when discussing topological exciton condensates (Seradjeh [2012]).

## 4.4. Conclusions

To sum up, we have considered several proposals for creating Majorana fermions at the ends of hybrid structures involving either an ExC or a BCS SC. In both cases we have shown that the low energy behavior of the system is identical to the Kitaev model. We have also discussed how to detect the emergence of these specific bound states within the BCS gap and have demonstrated that higher cumulants represent ideal tests for the true nature of the bound state. The conductance features alone may be ambiguous but the CGF is unique.

# Chapter 5

## Excitonic systems

In Section 4.3 of the last Chapter we have already adressed a new facet of ExCs. Indeed, even without Majorana fermions at the edges, ExCs allow for a rich variety of transport phenomena that we want to address in more detail in the first part of this Chapter. In Section 2.5 we describe the model developed in (Dolcini et al. [ 2010]) that we will use in order to calculate the CGF and discuss both the non-linear conductance and higher cumulants (Soller and Komnik [ 2013], Soller et al. [ 2012c]). Moreover, we shall take a mesoscopic view on drag-counterflow geometries where the top layer is contacted by leads at different chemical potentials inducing a current in the bottom layer that is also part of another circuit (Su and MacDonald [ 2008]). In our case we study a QPC and a quantum dot coupled to a phonon mode between the two leads of the bottom layer and explore the possibility of transforming current on the nanoscale.
In the second part of this Chapter we will concentrate on the interactions between excitons. The dipolar nature of excitons will allow us to investigate a hypothetical crystal phase and its detection using an applied laser field.

## 5.1. Charge transfer statistics of exciton condensates

We study a double layer ExC which is contacted by four metallic electrodes as illustrated in Fig. 5.1. While no inter-layer tunneling is assumed to occur, the two layers are coupled via Coulomb interaction which gives rise to the BCS condensation (of Kosterlitz-Thouless type) of the excitons. Following the description in Section 4.3 the Hamiltonian modeling the system reads

$$H = H_4 + H_{\mathrm{T,ExC}} + H_{\mathrm{ExC}}, \tag{5.1}$$

where the term $H_4$ accounts for the four metallic electrodes, characterized by electrochemical potentials $\mu_{\alpha\sigma}$, Fermi distribution functions $n_{\alpha\sigma}$, and an energy-independent DOS $\rho_0$. $\alpha = L, R$ refers to the contacts on the left/right side of the bilayer, whereas $\sigma = T, B$ labels the top and bottom layer, respectively. $H_{\mathrm{T,ExC}}$ describes the particle tunneling between the layers of the ExC and the

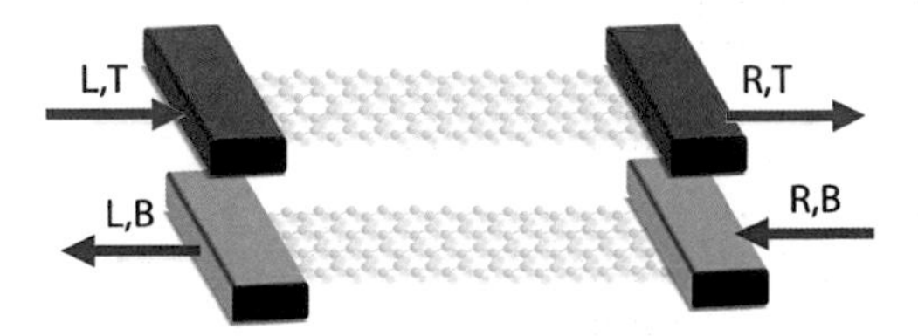


**Figure 5.1.:** Sketch of the experimental setup. The double layer ExC is contacted with four metallic electrodes.

metallic contacts

$$H_{\mathrm{T,ExC}} = \sum_{\sigma=T/B,\, \alpha=L/R} \gamma_{\alpha\sigma}(\alpha_\sigma^\dagger \Psi_\sigma + \Psi_\sigma^\dagger \alpha_\sigma)\,, \tag{5.2}$$

where $\gamma_{LT,B}$, $\gamma_{RT,B}$ are the tunneling amplitudes, $L_{T/B}$, $R_{T/B}$ the electron field operators for the four leads, and $\Psi_{T,B}$ the field operators for electrons in the ExC layers at the position $x = 0, l$ (for $L, R$), respectively. Additionally, we would need to sum over spin degrees of freedom. However, either spin is irrelevant since the system is a non-interacting mixture of ↑- and ↓-spins (Nozières [1982]) or (using a high enough magnetic field) the system is spin-polarized (Lozovik et al. [ 2002]). Therefore, we work with a spinless system and one may introduce the necessary prefactor for spin if necessary.
The Hamiltonian in Eq. (2.10) is reminiscent of the Bogoliubov-de Gennes Hamiltonian for a SC, where the layer indices $T$ and $B$ play the role of spin indices in BCS theory. Formally, the system has thus some analogies with a hybrid normal-SC-normal (NSN) junction. There are, however, two important differences. First, while the BCS order parameter couples electron (Cooper) pairs, the ExC order parameter couples electron-hole pairs, as it is clear from the last line of Eq. (2.10), where the additional transformation of holes in the bottom layer into electrons must be performed to recover the BCS Hamiltonian (Lozovik and Berman [ 1997]). Second, while in a SC Cooper pairs belong to the same physical system, in a bilayer geometry electrons and holes live on separate layers, which are characterized by different chemical potentials and which are independently contacted (Eisenstein et al. [ 1990]). As a consequence, this geometry allows to selectively tune the electrochemical potential of each 'spin' species and to probe 'spin'-resolved transport. Because of these differences, the present bilayer setup is much richer than the NSN junction or other SC hybrid structures considered before (see Chapter 3 ).
The major difference to simple SC hybrids is the fact that the ExC is contacted from two sides. Therefore, the calculation of the currents and higher cumulants in this hybrid structure represents an essentially self-consistent problem, where the external currents depend on the electrochemical potentials of the two layers and on the excitonic order parameter, which in turn adjust to ensure charge conservation and no inter-layer tunneling, thereby affecting the external currents themselves. In order to proceed, some assumptions are thus necessary. In view of possible implementations with graphene, we shall consider a linear Dirac cone spectrum $H_{\mathrm{Di}}$ for the layers, oppositely shifted by e. g. two external gates $\pm V_g$, so that $H_{T/B} = H_{\mathrm{Di}} \mp eV_g - \mu_{\mathrm{ExC,T/B}}$, where $\mu_{\mathrm{ExC,T/B}}$ are the electrochemical potentials. The length $l$ (see Fig. 5.1 ) is also assumed sufficiently large to neglect electromagnetic coupling between the two junctions, i.e. mutual capacitance between the normal wires, propagating modes in the ExC and similar effects as done for a superconducting wire in (Golubev and Zaikin [ 2010]). It is sensible to focus on the incoherent tunneling regime,$\xi_{\mathrm{ExC}}, l_\phi \ll l$, where $l_\phi$ is the dephasing length. The condition $\xi_{\mathrm{ExC}} \ll l$ also implies that self-consistency effects on the space dependence of $|\Delta(x)|$ are negligible (Su and MacDonald [ 2008]). A space-dependent

phase $\arg(\Delta(x)) \sim qx$, on the other hand, although essential to ensure that the ExC carries counter-flowing currents in the bulk of the bilayer, is not necessary for evaluating the currents in the leads, which are of interest here. In contrast, self-consistency of the electrochemical potentials $\mu_{\mathrm{ExC,T/B}}$ of the two layers is crucial to ensure current conservation in each layer (Lambert [ 1991])

$$\langle I_{LT} \rangle = \langle I_{RT} \rangle, \qquad \langle I_{LB} \rangle = \langle I_{RB} \rangle. \tag{5.3}$$

Under these assumptions, we have obtained the complete analytical expression for the CGF for all parameter regimes. Such expression, which has been used for our numerical evaluation, is quite lengthy and we do not report it here. Rather we use the approximation leading to Eq. (3.16) in the normal metal-SC case (see also Appendix A Eq. ( A.24), (A.25)) and the CGF acquires the following form on the left leads

$$\begin{aligned} \ln \chi_{\mathrm{ExC}}(\boldsymbol{\lambda})|_{\lambda_{R\sigma}=0} &= 2\tau \int \frac{d\omega}{2\pi} \\ \times \Bigg( \sum_{\sigma=T,\,B} & \ln \Big\{ 1 + T_\sigma(\omega) \Big[ (e^{i\lambda_{L\sigma}} - 1) n_{L\sigma}(1-f_\sigma) + (e^{-i\lambda_{L\sigma}} - 1) f_\sigma (1 - n_{L\sigma}) \Big] \Big\} \, \theta \left( \frac{|\omega_\sigma| - \Delta}{\Delta} \right) \\ + \ln & \Big\{ 1 + T_A(\omega) \Big[ (e^{i\lambda_{LT}} e^{-i\lambda_{LB}} - 1) n_{LT}(1-n_{LB}) + (e^{i\lambda_{LB}} e^{-i\lambda_{LT}} - 1) n_{LB}(1-n_{LT}) \Big] \Big\} \\ \times \theta & \left( \frac{\Delta - \max(|\omega_T|,\, |\omega_B|)}{\Delta} \right) \Bigg), \end{aligned} \tag{5.4}$$

where the transmission coefficients are given by $T_\sigma(\omega) = 4\tilde{\Gamma}_{L\sigma}/(1+\tilde{\Gamma}_{L\sigma})^2$ and $T_A(\omega) = 4\tilde{\Gamma}_A/(1+\tilde{\Gamma}_A)^2$. The effective transparencies are parametrised by the ExC DOS as $\tilde{\Gamma}_{L\sigma} = \Gamma_{L\sigma}|\omega_\sigma|/\sqrt{\omega_\sigma^2 - \Delta^2}$ and $\tilde{\Gamma}_A = \Gamma_{LT}\Gamma_{LB}\Delta^2/[\sqrt{\Delta^2 - \omega_T^2}\sqrt{\Delta^2 - \omega_B^2}]$, where $\Gamma_{L\sigma} = \pi^2 \rho_{0L\sigma}\rho_{0E}\gamma_{L\sigma}^2/2$. The functions $f_T$ and $f_B$ denote Fermi distributions for the quasiparticles in the separate layers and $\omega_{T,B} = \omega - \mu_{\mathrm{ExC,T/B}}$.

The first line of Eq. (5.4) describes the supra-gap contribution, which is only due to single electron transport and is characterized by the normal transmission coefficient $T_\sigma$. In contrast, the second line describes the sub-gap contribution due to the phenomenon of excitonic AR (Su and MacDonald [2008]), consisting of an electron and a hole (traveling in different layers), which enter or leave coherently the bilayer in order for an excitonic pair to be transferred along the bulk of the system.

The respective currents may be obtained as the first derivative of Eq. (5.4). Imposing the self-consistency condition (5.3) determines $\mu_{\mathrm{ExC}\sigma}$, and one obtains the final results for the two currents $I_\sigma \doteq \langle I_{L\sigma} \rangle = \langle I_{R\sigma} \rangle$. For simplicity, we consider the symmetric junction case, $\Gamma_{LT} = \Gamma_{RT}$ and $\Gamma_{LB} = \Gamma_{RB}$, and symmetrically applied biases $\mu_{\mathrm{LT}} = -\mu_{\mathrm{RT}} = -V_T/2$, $\mu_{\mathrm{LB}} = -\mu_{\mathrm{RB}} = -V_B/2$. In this case Eq. (5.3) is always fulfilled for $\mu_{\mathrm{ExCT}} = \mu_{\mathrm{ExCB}} = 0$.

The average currents are plotted in Fig. 5.2 (a) as a function of the top layer bias$V_T$, for a fixed value of the bottom layer bias $V_B$. As one can see, because of the EC coupling, both $I_T$ and $I_B$ change, even when varying $V_T$ only. In particular, for $|V_T|, |V_B| < 2\Delta$, one observes $I_T = -I_B$, a signature that in the sub-gap regime transport can only occur via excitonic counterpropagating currents in the bulk of the layers, which are transformed into electron and hole currents in the leads through excitonic AR. Notice that for the value $V_T = V_B$ a current locking occurs ($I_T = I_B = 0$), because the ExC cannot sustain currents driven by equally applied biases (exciton blockade). At $V_T = 2\Delta$ excitonic pairs start to break up and the resulting electrons/holes get excited above the gap. This is clearly shown in Fig. 5.2 (b), where the positive conductance exhibits a resonance peak, whereas the negative transconductance abruptly changes sign. At higher voltage values the ExC plays a minor role, so that the conductance tends to the value of the case $\Delta = 0$, and the transconductance vanishes, indicating that transport in the bottom layer is independent of the

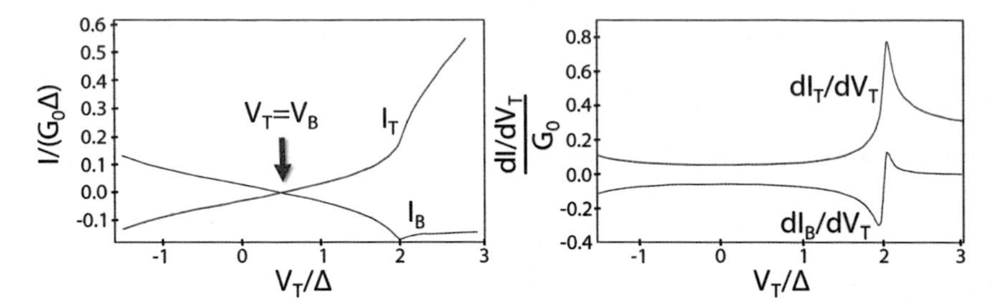


**Figure 5.2.:** **(a)**: Current in the top (red) and bottom (blue) layer as a function of $V_T$, for a fixed value $V_B = \Delta/2$. $\Gamma_{LT} = \Gamma_{LB} = 0.171$ corresponds to transmission of 0.5 for the uncoupled system, $T = 0.01\Delta$. For $|V_T|, |V_B| < 2\Delta$ the bilayer exhibits counterpropagating currents, exciton blockade occurrs at $V_B = V_T$.
**(b)**: Differential conductance $dI_T/dV_T$ (red) shows a resonance peak and tends to the typical value for a QPC whereas the transconductance $dI_B/dV_T$ shows a resonance peak before vanishing at larger bias.

voltage applied to the top layer.
So far experiments on transport through ExCs have only been made in a typical drag configuration ($V_B = 0$). The inset of Fig. 5.3 shows a sketch of the experiment described in Nandi et al. [ 2012], where two Corbino disks are confined in a GaAs/AlGaAs double quantum well structure. The experiment is performed in the $\nu = 1$ quantum Hall state of the system and we expect exciton condensation between the two Corbino disks that serve as the top and bottom layer indicated in Fig. 5.1 . In the top layer a current can be excited and additionally the induced drag current in the bottom layer is recorded.
We calculate the current from Eq. (5.4) and use $\Gamma_T = 0.02$, $\Gamma_B = 0.72$, $\Delta = 100\,\mu$eV. We observe good qualitative agreement in Fig. 5.3 . We only observe a slight deviation in the bottom current for voltages above the ExC gap. We ascribe this discrepancy to a small interlayer coupling that might still be present in the system and is also suggested from the analysis of the experimental data (Nandi et al. [ 2012]).
The current correlators can be obtained from Eq. (5.4) as

$$\langle\langle I_{\alpha\sigma} I_{\alpha'\sigma'} \rangle\rangle = (-i)^2 \frac{1}{\tau} \frac{\partial^2}{\partial\lambda_{\alpha\sigma}\partial\lambda_{\alpha'\sigma'}} \ln \chi \Big|_{\lambda_{\alpha\sigma}=0=\lambda_{\alpha'\sigma'}} . \tag{5.5}$$

In equilibrium ($V_T = V_B = 0$) we obtain the customary Johnson-Nyquist relation for $T \ll \Delta$

$$\langle\langle I_{\mathrm{LT}} I_{\mathrm{LT}} \rangle\rangle|_{\mathrm{eq}} = -\langle\langle I_{\mathrm{LT}} \mathrm{I}_{\mathrm{RT}} \rangle\rangle|_{\mathrm{eq}} = 4T_A(0) G_0 k_B T. \tag{5.6}$$

This result indicates that the two electrons involved in an excitonic AR dwell in separate layers so that only the conductance of a single layer enters the Johnson-Nyquist noise. The second equality in (5.6) is the generalized Johnson-Nyquist relationship obtained in Anantram and Datta [ 1996] for a floating superconductor, which is here obtained without any use of Langevin forces.
Likewise shot noise in the tunneling limit ($\Gamma_{LT}, \Gamma_{LB} \ll 1$) follows the Schottky formula for single electrons

$$S = (-i)^2 \frac{1}{\tau} \frac{\partial^2}{\partial\lambda_{L\uparrow}^2} \ln \chi_L(\lambda_{L\uparrow}, \lambda_{L\downarrow}) \Big|_{\lambda_{\alpha\sigma}=0} = e\langle\langle I_{L\uparrow} \rangle\rangle, \tag{5.7}$$

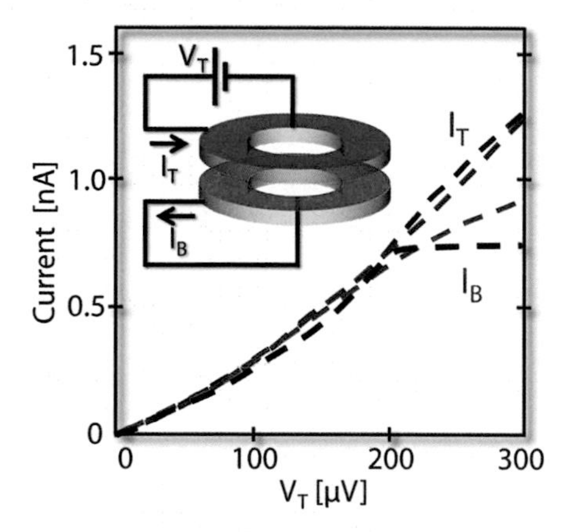


**Figure 5.3.:** Comparison of the prediction for the current using Eq. (5.4) for a typical drag experiment ($V_B = 0$) in ExCs. We use fitting parameters $\Gamma_T = 0.02$, $\Gamma_B = 0.72$, $\Delta = 100\,\mu$eV and take the temperature $T = 17$ mK from the experiment. The dashed blue and red curve correspond to the top and bottom current data from the experiment, respectively and the dashed black curves are the results from our model.

and the third cumulant follows the Levitov-Reznikov formula (Levitov and Reznikov [ 2004])

$$C_3 = \frac{i}{\tau}\frac{\partial^3}{\partial\lambda_{L\uparrow}^3}\ln\chi_L(\lambda_{L\uparrow},\lambda_{L\downarrow})\Big|_{\lambda_{L\downarrow}=\lambda_{L\uparrow}=T=0} = e^2\langle\langle I_{L\uparrow}\rangle\rangle. \tag{5.8}$$

As we see from Fig. 5.4 also in non-equilibrium one always observes a negative cross-correlation. This is different from the case of a SC contacted to two normal electrodes, where one observes a positive cross-correlation of the two currents in the normal leads via crossed AR (Recher et al. [2001]). We will discuss this phenomenon in more detail in Chapter 6 . The reason for this crucial difference is the fact that in the ExC case we probe the correlation of electrons and holes rather than correlations of electron pairs as in the case of SCs.
We call the two voltage bias situations $V_T = \pm V_B$ parallel and antiparallel configurations. In the first configuration, where the average current is vanishing, the noise and the cross-correlation are shown in Fig. 5.4 (a). In the sub-gap regime, up to thermal fluctuation effects, both the noise and the cross-correlation vanish. This is because the incoming electrons and holes are always reflected back into the same lead they are injected from, since no exciton can penetrate inside the ExC. Notice that this effect is essentially independent of the interface transmission, as shown in the three curves of Fig. 5.4 (a). Indeed, for the parallel bias configuration in the subgap regime the ExC gap effectively plays the role of a large barrier. In contrast, when $|V| > 2\Delta$, quasiparticles can be excited above the gap, and the noise in each lead starts deviating from zero, eventually increasing linearly with $V$. In the antiparallel configuration, where the average currents in the layers flow in opposite directions, the noise and the cross correlation are shown in Fig. 5.4 (b) as a function of$V = V_T = -V_B$. In the supra-gap regime the behavior is qualitatively similar to the parallel configuration, so that the noise increases and the cross-correlations saturate to a value determined by the quasiparticle mixed character above the gap. However, differences with respect to the parallel configuration emerge in the sub-gap regime, where both noise and cross correlation are now non-vanishing, and depend on the interface transmission.

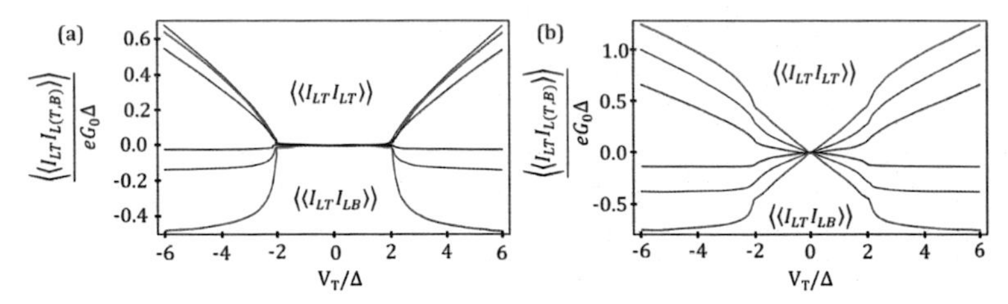


**Figure 5.4.: (a)**: The parallel bias configuration (exciton blockade). Noise and cross-correlations are shown as a function of $V_T = V_B = V$, for the three values of transmission for the uncoupled system of 0.3 (black), 0.5 (blue), 0.7 (red) and $T = 0.01\Delta$.
**(b)**: The antiparallel bias configuration (counterpropagating currents). Noise and cross-correlations are shown as a function of $V_T = -V_B = V$, again for the three values of contact transparency.

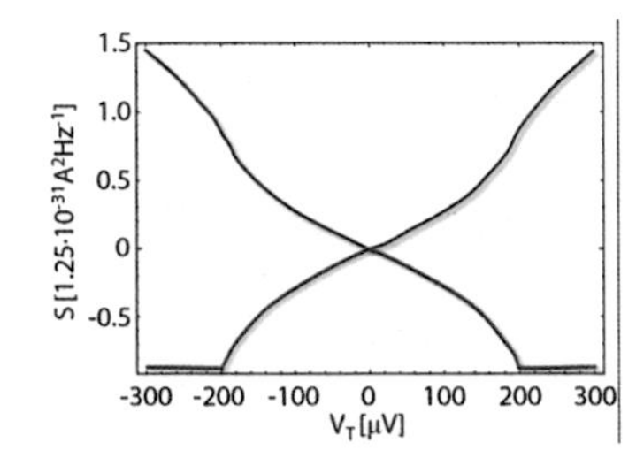


**Figure 5.5.:** Noise for the top current (upper blue curve) and the cross-correlation of the top and bottom current (lower red curve) using Eq. (5.4) and fitting parameters $\Gamma_T = 0.02$, $\Gamma_B = 0.72$, $\Delta = 100\mu$eV and $T = 17$ mK for the experiment (Nandi et al. [ 2012]).

The drag configuration ($V_B = 0$) is qualitatively similar to the antiparallel bias configuration. For this case we analyse the noise and cross correlation of the top and bottom current for the setup discussed in Fig. 5.3 . We use the same parameters as used for the explanation of the current features and obtain the result depicted in Fig. 5.5 .
Again, we observe finite noise of the top current and also a negative cross-correlation.

## 5.2. Nanotransformation

One possible application of ExCs could be as nanoscale voltage transformers. Ideal voltage transformers, such as inductors, require that the transformation coefficient depends weakly on the load characteristics, and that energy losses are minimal. However, implementation of on-chip silicon based inductors turns out to be very difficult (Cerofolini [ 2011]), so that the seek for trade-off solutions in nanotransformers represents a great challenge in modern electronics.
The contact configuration proposed in (Su and MacDonald [ 2008]) suggests that the setup can be used as a voltage transformer at the nanoscale, contacting the ExC bottom layer to another mesoscopic system with a known $I - V$ characteristics, such as a QPC, where $I_{\text{QPC}} = G_0 T_1 V_B$, with $T_1$ being the contact transparency (see Fig. 5.6 (a)). At low temperatures and for $|V_{T,\,B}| < 2\Delta$,

only excitonic ARs contribute to transport and determine $\langle I_{\rm LT}\rangle$ as a function of $V_T$. On the other hand $\langle I_{\rm LT}\rangle = -\langle I_{\rm LB}\rangle$, and current conservation implies that $-\langle I_{\rm LB}\rangle = I_{\rm QPC}$. This leads to the transformation

$$\langle I_{\rm LT}\rangle(V_T) = G_0 T_1 V_B. \tag{5.9}$$

A typical example for the voltage interrelation is shown in Fig. 5.6 (b). In this case the transformation is controlled by $T_1$, which typically is tunable. Importantly, nanotransformers based on the dissipationless EC counterflowing currents may help minimizing heat and noise production.

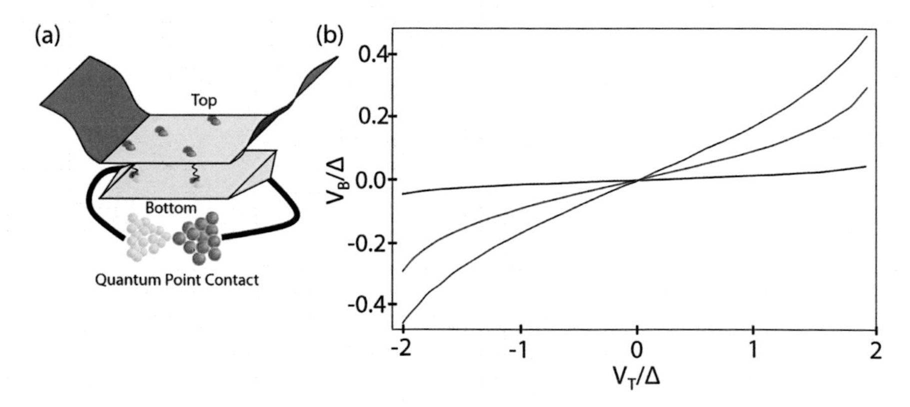


**Figure 5.6.: (a)**: Sketch of the experimental setup for transformation of voltages using the ExC.
**(b)**: We show $V_B$ due to the current in the bottom layer coupled to a QPC as a function of the applied voltage $V_T$ for fixed $\Gamma_{LT} = 0.2 = \Gamma_{RT} = \Gamma_{LB} = \Gamma_{RB}$ and the transparency of the QPC is varied from $T_1 = 0.1$ (blue curve), $T_1 = 0.3$ (red) to $T_1 = 0.7$ (black).

The principle of this transformer can also be used in contact with an actual transistor, demonstrating its applicability in typical electronic circuits. We use a transistor based on coupling to internal degrees of freedom, see Fig. 5.7 .
In this case we connect the ExC to a quantum dot which in turn is coupled to a local phonon mode.

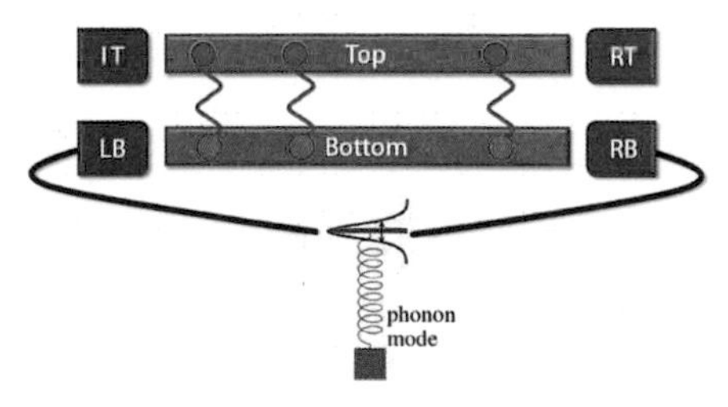


**Figure 5.7.:** Sketch of the setup for transformation of current using an ExC and a nanoscale transistor: the bottom layer is connected to a quantum dot which in turn is coupled to a phonon mode.

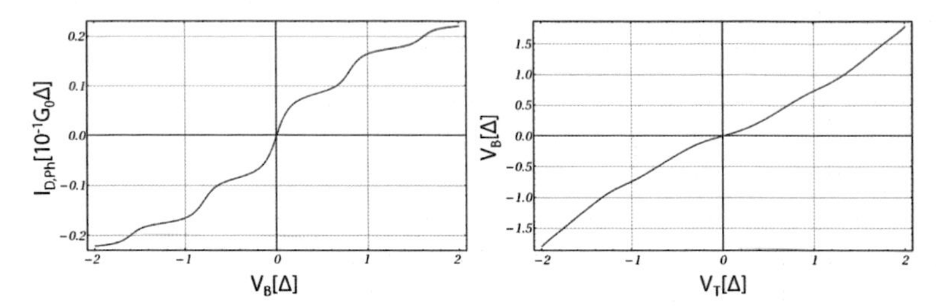


**Figure 5.8.:** Transformation of voltages using the ExC coupled to a quantum dot in turn coupled to internal degrees of freedom. The left panel shows the current voltage characteristics $I_{\mathrm{D,Ph}}$ calculated from (Kast et al. [ 2011]) for$\Gamma_L = \Gamma_R = 0.05\Delta$. Additionally, we used $\omega_0 = 0.4\Delta$, $M_0 = 1.6\Delta$, $\epsilon_D = 0$ and $\beta = 1/T = 20\Gamma_L$. In the right panel we show the voltage $V_B$ of the induced current in the lower layer coupled to the quantum dot as a function of the applied voltage $V_T$ in the upper layer for the same set of parameters. The couplings have been chosen to be $\Gamma_{LT} = 0.3 = \Gamma_{RT} = \Gamma_{LB} = \Gamma_{RB}$. All voltages have been rescaled with $\Delta$. One observes the current steps in both the transformation relation and the $I - V$ characteristics of the quantum dot. However, there is a strong rectifying trend from the ExC.

The latter system is described by the Hamiltonian (Anderson-Holstein model), which we will discuss in more detail in Chapter 9

$$\begin{aligned} H_{\mathrm{AH}} &= H_{\mathrm{LR}} + H^{(I)}_{\mathrm{D,LR}} + H_D + H^{(I)}_{\mathrm{D,Ph}} + H_{\mathrm{Ph}} \\ &= \sum_{k=L,R} \epsilon_k a_k^+ a_k + \gamma \sum_{k=L,R} (a_k^+ d + \mathrm{H.c.}) + \epsilon_D d^+ d + d^+ d M_0 (b^+ b) + \omega_0 (b^+ b + 1/2), \end{aligned}$$

where $\epsilon_D$ is the position of the dot level, $M_0$ is the phonon coupling strength and $\omega_0$ is the resonance frequency of the phonon mode. The tunnel rates to the dot are given by $\Gamma_{L,R} = 2\pi\rho_{0L,R}\gamma^2$ with the DOS of the leads $\rho_{0L,R}$.
Since we want to describe this setup in the presence of strong phonon interaction in order to operate it as a transistor, a perturbative approach cannot be used. So far, exact approaches to the problem have been achieved only by means of diagrammatic Monte Carlo (Mühlbacher and Rabani [2008]) and the multilayer multiconfiguration time-dependent Hartree method in second quantization representation (Albrecht et al. [ 2012b]). Analytical approximations (Maier et al. [ 2011]) and a rate equation approach (Kast et al. [ 2011]), however, have been shown to yield good quantitative agreement. We use the rate equation approach of Kast et al. [ 2011] to obtain the current-voltage characteristics $\langle I_{\mathrm{D,Ph}}\rangle$ of the quantum dot. Using such expression we can use the above argument leading to Eq. (5.9)

$$\langle I_{LT}\rangle(V_T) = \langle I_{\mathrm{D,Ph}}\rangle. \tag{5.10}$$

The result is depicted in Fig. 5.8 , where we show that the voltage can indeed be transformed and depends on the current-voltage characteristics of the quantum dot. The $I - V$ characteristics of the quantum dot show characteristic steps that are associated to the possibility of exciting phonons at the frequency $\omega_0$. Although the phonon-induced current steps are clearly visible in the $I - V$, there is a strong rectifying trend from the ExC when it comes to the transformation relation.
We have therefore demonstrated the possbility of transforming current on the nanoscale in two different proposals.

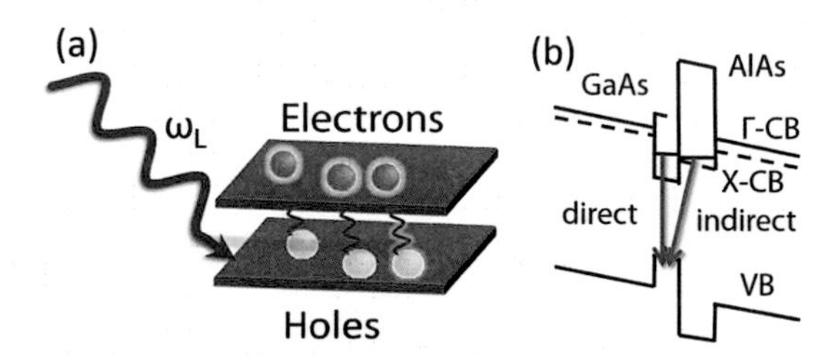


**Figure 5.9.: (a)**: Shows the typical experimental situation: excitons are created in a coupled quantum well via an applied laser field that knocks out electrons from the valence band into the conduction band and thus forms excitons.
**(b)**: Possible exciton states in a typical coupled quantum well such as GaAs/AlAs. These coupled quantum wells have two conduction bands ($\Gamma$ and $X$) originating from different points in the Brillouin zone and allow for both the formation of direct and indirect excitons, however, with different binding energies.

## 5.3. Exciton crystals

In this Section we move away from the BCS condensation of excitons and study another essential feature of a system of interwell excitons which are the parallel electric dipole moments which become important at intermediate exciton densities where condensation is prohibited but the excitons still 'feel' their mutual interaction. In this case the concept of an exciton crystal emerged (Kulakovskii et al. [ 2004]). We will follow the discussion in Breyel et al. [ 2013].
A typical measurement would start via a laser pulse in a coupled quantum well structure (e.g. GaAs/AlAs). Afterwards, measurements are performed during which the excitons are eventually deexcited and afterwards the system is ready for another preparation (Butov et al. [ 1994], Gärtner et al. [ 2007]).
We assume that in every measurement cycle the system consists of $N$ electrons, located at random positions $\mathbf{r}_i$. We excite the system via a laser field at a frequency $\omega_L$. $\Delta$ represents the Rabi frequency and $\Delta_d$ describes the detuning of the laser field from this transition. In a typical experiment one uses short and high intensity laser pulses where the Rabi frequency is of the order of the exciton binding energy (Östreich and Knorr [ 1993]). A cicularly polarized light beam (polarization$\sigma$) at a suitably chosen frequency can create direct and indirect neutral excitons with well defined spins $\sigma$ and $\bar{\sigma} = -\sigma$ via exciting an electron from the valence band of the same or the other semiconductor in the coupled quantum well (Türeci et al. [ 2011]), see Fig. 5.9 .
In the following we will focus only on the indirect excitons. Direct excitons can be neglected either by an appropriate choice of the excitation frequency $\omega_L$ or by simply waiting long enough since their recombination time is much shorter. The indirect excitons possess dipole moments $d$ perpendicular to the layers $d = eD$, where $D$ is the interlayer separation, see Fig. 2.1 (a).
Additionally, we want to focus on the major effects that arise due to exciton-exciton interaction. We treat the exciton as a two level system (Kowalik-Seidl et al. [ 2010]) with electrons either in the conduction band or excited to an exciton (Astrakharchik et al. [ 2007]). We assume the levels to be sharp neglecting effects from the Fermi distribution of the separate bands. This is justified in the limit of a large detuning of the laser from the resonance (Östreich and Knorr [ 1993]). The electrons interact with the laser light and with each other through the dipole interaction when they form excitons. The velocity distribution of excitons in coupled quantum wells (Snoke et al. [ 2002]) can

be tuned by efficient cooling (High et al. [ 2012]) and the application of a perpendicular magnetic field, which gives rise to a higher effective mass of the excitons (Lozovik et al. [ 2002]). With both the laser pulse duration (Vörös et al. [ 2005]) and the experimental measurement taking$\tau < 0.1$ ns time and the resulting velocities $v_{\mathrm{Ex}} \approx 10$ m/s the displacement of a single exciton is $v_{Ex}\tau < 1$ nm. The typical separations of the excitons we will encounter are of the order 10 nm so that we can assume the excitons to be fixed in space. Since the electrostatic properties do not depend on the details of exciton states we model the system as a randomly arranged interacting ensemble of spin-1/2 sub-systems each representing a single electron/exciton. Hence, the Hamiltonian reads (Astrakharchik et al. [ 2007], Breyel et al. [ 2013], Robicheaux and Hernández [ 2005])

$$H_{\mathrm{crystal}} = -\frac{\Delta_d}{2}\sum_{i=1}^{N}\sigma_z^{(i)} + \frac{\Delta}{2}\sum_{i=1}^{N}\sigma_x^{(i)} + \frac{C_{\mathrm{dd}}}{4}\sum_{i=2}^{N}\sum_{j=1}^{i-1}\frac{(1+\sigma_z^{(i)})(1+\sigma_z^{(j)})}{|\mathbf{r}_i-\mathbf{r}_j|^3}, \tag{5.11}$$

where $\sigma_{x,z}^{(i)}$ denote the Pauli matrices. This equation uses the rotating wave approximation when describing the light-matter interaction as in Lindberg and Koch [ 1988].
This can be interpreted as a (generalised) spin-1/2 Ising model. The transition frequency $\Delta$ can be interpreted in the spin language as a magnetic field perpendicular to the quantization axis, which we choose to be the $z$-axis. The detuning $\Delta_d$ corresponds to a magnetic field in $z$-direction. The third parameter $C_{\mathrm{dd}}$ indicates the strength of the effective interaction between the excitons. We should note that a similar model has originally been applied to interactions between Rydberg atoms, where a similar crystallization to a Wigner-crystal-like phase exists (Robicheaux and Hernández [ 2005]). We also note that we are only interested in the case $\Delta_d > 0$, because otherwise it is energetically not favorable to excite excitons. This allows us to use $\Delta_d$ as our basic energy scale.
We typically use a large detuning from the transition frequency in accordance with experimental studies (Snoke et al. [ 2002]) so that$\Delta_d$ is comparable to the exciting laser field. A typical laser field has an exitation frequency of 1 eV. The excitons can be approximated as bosons with the dipolar moment oriented perpendicular to the plane. In this case $C_{\mathrm{dd}} = e^2D^2/\epsilon$. For the dielectric spacer between the top and bottom layer we assume $\epsilon = 3.9\epsilon_0$ being a typical value for $SiO_2$ and $D = 100$ nm. The interaction can be described by a dimensionless parameter $C_{\mathrm{dd}}/(\Delta_d L^3)$ and we take the length $L$ of the simulated square in 2D to be $\approx 8$ nm.
Just as in the actual experiments we calculate averages over a large number of different, randomly sampled electron arrangements in the bilayer. In each such cycle we solve the Hamiltonian in Eq. (5.11) for $N$ electrons being playced at random positions. The ground state (GS) is found by means of exact diagonalisation or an approximative relative of it and from the GS we determine the number of excitons. Repeating this routine numerous times generates statistics from which properties of the system can be deduced. The number of excitons may be non-integer as we take the average over many samples.
After having distributed the electrons randomly in a given volume (one- or two-dimensional) the corresponding Hamiltonian matrix is calculated in the basis in which every electron is represented by a single spin. In the case of the approximative version (for large $N$) the size of this matrix is reduced by the truncation of the Hilbert space which is done by only taking into account the $N_{\mathrm{trunc}}$ basis states with the smallest diagonal elements in the Hamiltonian matrix. We systematically checked that all the results presented do not rely on $N_{\mathrm{trunc}}$ which is true as long as $N_{\mathrm{trunc}}$ is larger than a certain threshold. Now the eigenvector corresponding to the smallest eigenvalue (the GS) is calculated. It is given as a linear combination of the previously mentioned basis states and therefore enables us to easily evaluate different observables. Furthermore we calculate the pair correlation

function

$$g(\mathbf{r}) = \langle \rho_{\text{Exciton}}(\mathbf{r})\rho_{\text{Exciton}}(0)\rangle, \tag{5.12}$$

where $\rho_{\text{Exciton}}$ is the number of excitons in a volume element $d\mathbf{r}$ around $\mathbf{r}$. However, due to rotational symmetry we may additionally average $g(\mathbf{r})$ over the polar angle and arrive at the relevant quantity $g(r)$ which only depends on the relative distance. We compute it in the following way: we divide the possible range for distances between electrons in our system into equidistant bins and measure the distance between each pair of electrons to assign it to a certain bin. For each pair the squares of coefficients (from the representation of the GS as a linear combination) are summed over those states in which the particular pair of electrons is excited. The sum over multiple random arrangements then produces the correlation function. Details of the implementation may also be found in (Breyel et al. [ 2012]).

We first discuss the properties of the system in 2D: for weak interaction the excitons are evenly distributed, whereas for strong interaction they tend to populate the boundaries. However, for an appropriate intermediate choice of the interaction $C_{\text{dd}}$ a sizeable fraction of the excitons is redistributed to the center of the 2D plane. In this case we calculate the density distribution of excitons for a square of size $(8\,\text{nm})^2$ and observe the onset of the typical regular ordering of Wigner-crystal type, see Fig. 5.10 .

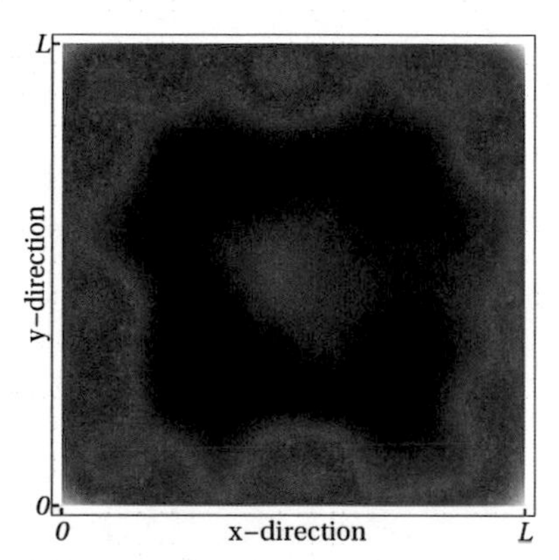


**Figure 5.10.:** Density distribution of excitons. Bright color indicates high density, dark color low density. One can see that the edges of the 2D volume are preferred over the inside. An additional peak, which is not as pronounced as the edges, can be found at the center. The plot was taken at $\Delta_d = 1$, $\Delta = 0.1$, $C_{\text{dd}} = 125$ and $L = 10$, which results in realisations with 5 and 6 excitons being most probable.

The emerging long-range order is even more apparent when considering $g(r)$ for a similar set of parameters, see Fig. 5.11 . In order to exclude possible finite-size effects for its calculation we switch to periodic boundary conditions. We immediately identify a blockade radius $R_B$ as the position of the first maximum. It is the distance between two excitons up to which it is disadvantageous for both electrons to form an exciton. Beyond the $R_B$ the quantitative behavior of $g(r)$ strongly depends on $C_{\text{dd}}$, but additional peaks in Fig. 5.11 are clearly visible. In an ideal crystal these peaks would be sharp but in our simulation this cannot be achieved due to the limited number of electrons considered.

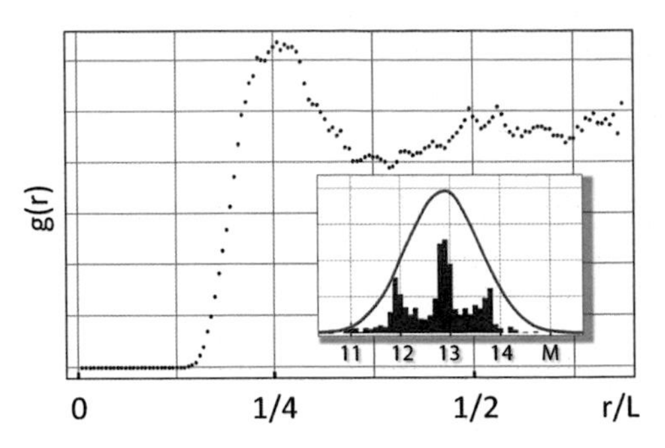


**Figure 5.11.:** Pair correlation function at $\Delta_d = 1$, $\Delta = 0.1$, $C_{\rm dd} = 8$, and $L = 10$. The blockade radius as well as the first two maxima of the curve are clearly visible. The double peak of the second maximum really is a single peak and only splits 'by coincidence' in the sample shown. The correlation function is measured in arbitrary units. If needed it can be normalized to the large distance limit. The data points for the largest distances are less reliable than the rest since only very few runs contribute to these points. The inset shows the histogram of the number of excitons with a Gaussian fit for the same parameters as for $g(r)$.

The blockade phenomenon is a clear feature to be searched for in future experiments of the type described in (High et al. [ 2012]) since it clearly distinguishes the exciton crystal-like and liquid-like behaviour from the BEC condensate where one observes a peak in $g(r)$ at $r = 0$.
The inset of Fig. 5.11 also shows the histogram$P(M)$, generated by considering all arrangements for the excitons. For $\Delta = 0$ the number-operator for excitons commutes with the Hamiltonian in Eq. (5.11) and one obtains a set of discrete peaks at integer values $M$. For the case of $0 < \Delta \ll \Delta_d$ considered here these peaks broaden up but the distribution remains almost discrete.

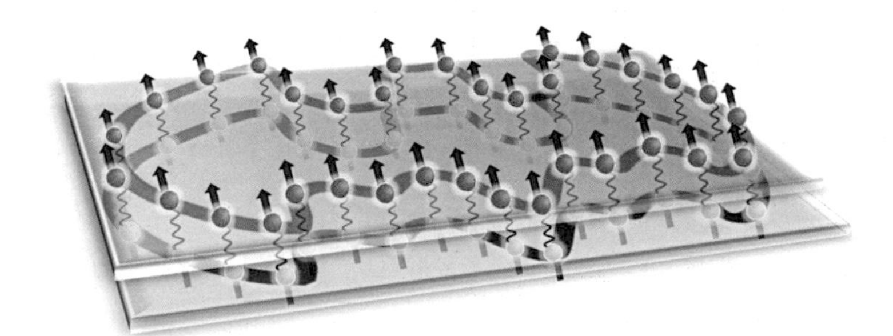

**Figure 5.12.:** Possible realisation of the crystalline phase of interacting exciton dipoles in a typical 1D confiment. The arrows illustrate the dipole moments and wiggly lines indicate the binding of excitons.

We also consider the case of an additional electrostatic confinement for the excitons as discussed

e.g. in Gärtner et al. [ 2007], so that the excitons have to be treated in 1D. A possible realisation of the crystalline phase in this case for a typical 1D confinement is shown in Fig. 5.12 .
For the pair correlation function and the histogram we find comparable results to those illustrated in Fig. 5.11 . We further investigated the average number of excitons$M$ as a function of the interaction strength $C_{\mathrm{dd}}$ as depicted in Fig. 5.13 . The inset of Fig. 5.13 also shows the standard deviation$\sigma_{\mathrm{Ex}}$ for a series of simulations. The change of the power law decay of $M$ is not as abrupt as in another possible realisation of a Wigner-crystal (Breyel et al. [ 2012]) but the standard deviation$\sigma_{\mathrm{Ex}}$ shows a clear change at $C_{\mathrm{dd}}/(\Delta_d L^3) \approx 10^{-3}$. We find this change to be heralding a phase transition in the system. The remarkable change of the variance as a function of $C_{\mathrm{dd}}$ shows that also simple

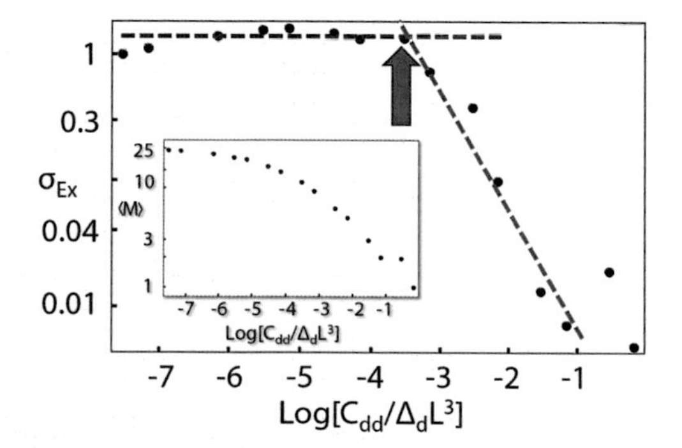


**Figure 5.13.:** Standard deviation of the mean number of excitons for $\Delta_d = 1$, $\Delta = 0.1$, and $L = 10$. The calculation was performed with $N = 25$ electrons. The straight lines and the arrow illustrate the phase transition. The inset shows the mean number of excitons, which is bounded by 25 from above due to our choice of $N$.

statistical analysis of the presence of excitons allows to identify the phase transition to an exciton crystal.

## 5.4. Conclusions

In conclusion, we have discussed two possible phases of excitonic systems: the BCS condensate and the exciton crystal. In the first case we have discussed current transport and higher cumulants of the charge transfer statistics. The most important findings include the possibility of perfect Coulomb drag via excitonic AR and a negative cross-correlation of currents equal in magnitude to the noise. We have compared our results to experimental data and presented possible applications involving the specific properties of current flow in ExCs. For the exciton crystal we discussed how to access the crystalline phase and its detection via the correlation function.

# Chapter 6

## Single dot Cooper pair splitters

In Section 2.2 we introduced the ground state wavefunction of a Cooper pair in Eq. ( 2.8). If, by some mechanism, we were able to separate the two electrons to go to two spatially separated observers $A$ and $B$ (typically referring to Alice and Bob) the resulting wavefunction would be given by

$$\Psi^- = \frac{1}{\sqrt{2}}(|\uparrow\rangle_A|\downarrow\rangle_B - |\downarrow\rangle_A|\uparrow\rangle_B), \tag{6.1}$$

which we cannot write as a product state $|\phi\rangle_A|\psi\rangle_B$. Such states are known as entangled states (Nielsen and Chuang [ 2000]). Indeed, the state in Eq. (6.1) is even maximally entangled, since measuring the spin on one of the electrons immediately determines the spin of the second electron. Such pairs of particles, also known as Einstein-Podolsky-Rosen (EPR) (Einstein et al. [ 1935]) pairs, therefore provide a valuable resource of quantum mechanics.
Two steps should be taken from here: first, one should provide a mechanism for separating electrons and secondly one should provide a measurement scheme to test the entanglement. In this Chapter we will concentrate on the first step using a basic setup with just a single quantum dot (QD) in order to investigate the basics of Cooper pair splitting. Chapter 7 will provide the connection to contemporary experiments. The second step will be investigated in Chapter 10 .

## 6.1. Non-interacting case

A basic splitting device is a three-terminal system with a central SC lead in between two normal ones (N1 and N2), as depicted in Fig. 6.1 . When a Cooper pair is injected from the SC it can either be transferred to one lead locally (AR) or its electrons can be transmitted to different leads, which corresponds to crossed AR (CAR) (Recher et al. [ 2001]) or Cooper pair splitting (CPS) (Hofstetter et al. [ 2011]).
Leaving both normal drains at the same chemical potential one can study the competition between

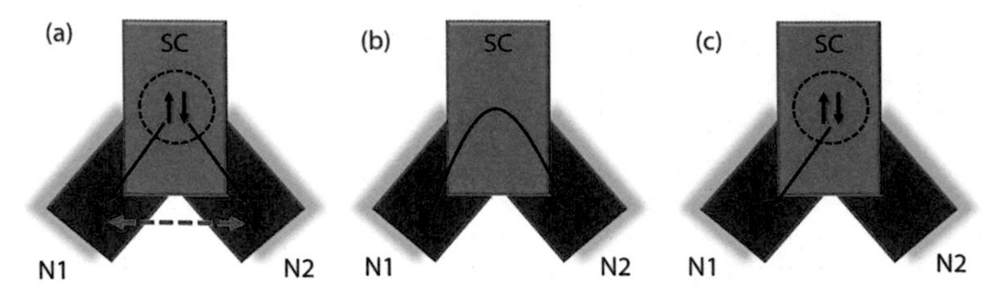


**Figure 6.1.: (a)**: Illustration of Cooper pair splitting in a Y-junction geometry with the SC (S) acting as a source for entangled electrons in the drains N1 and N2.
**(b)**: Process of elastic cotunneling that becomes possible at finite bias between N1 and N2.
**(c)**: Local AR transfers a pair of electrons to the same lead.

local processes like AR with the nonlocal process of CPS. However, if one applies a finite bias between the two normal leads the key issue is the competition between CPS and elastic cotunneling (EC) in the nonlocal conductance measured between N1 and N2.
Recently, the nonlocal conductance in SC hybrid structures has been extensively studied theoretically using different approaches (Brinkman and Golubov [ 2006], Falci et al. [ 2001], Recher et al. [2001]). Among other quantities the cross-correlation noise has been found to be an efficient tool for discrimination of the current caused by CPS from other current contributions (Bouchiat et al. [2003], Lesovik et al. [ 2001]) and it is directly accessible via the CGF (Börlin et al. [ 2002]).
Even in its basic realization a non-interacting QD usually shows a highly non-linear transmission leading to non-linear current-voltage relations. Combined with a SC electrode, the single-particle density of states of which is itself highly non-linear due to the finite gap, the resulting structure is expected to possess transport properties with a non-trivial voltage dependence. Thus far, these aspects have only been taken into account in Soller and Komnik [ 2011b] so that we will follow its argumentation.
The Hamiltonian for the system under consideration is given by

$$H_{\mathrm{SD}} = H_l + H_D + H_{\mathrm{TD}} + H_S. \tag{6.2}$$

The term $H_l$ describes normal or FM electrodes in the language of the respective electron field operators $\Psi_{\alpha\sigma}(x)$, where $\alpha = 1, \dots, N$ numbers the electrodes. The local density of states is $\rho_{0\alpha}$. Electron exchange between the QD and the electrodes is given by

$$\begin{aligned} H_{\mathrm{TD}} &= \sum_\alpha H_{\mathrm{TD}\alpha} + H_{\mathrm{TS}} && (6.3)\\ &= \sum_{\alpha,\sigma} \gamma_\alpha \left[d_\sigma^+ \Psi_{\alpha\sigma}(x=0) + \mathrm{H.c.}\right] + \sum_\sigma \gamma_S \left[d_\sigma^+ c_\sigma(x=0) + \mathrm{H.c.}\right], && (6.4) \end{aligned}$$

where $\gamma_\alpha$ or $\gamma_S$ are the tunneling amplitudes between the QD and the normal/FM or SC electrodes. $d_\sigma$ is the annihilation operator of an electron with spin $\sigma$ on the QD. The tunneling Hamiltonian is therefore a straightforward generalisation of Eq. (2.3), where tunneling between different leads does not occur directly but via the QD. The QD in the presence of a magnetic field $B$ along the polarisation-direction of the FMs is modelled by

$$H_D = \sum_\sigma (\epsilon_D + \sigma h/2) d_\sigma^+ d_\sigma =: \sum_\sigma \epsilon_{D\sigma} d_\sigma^+ d_\sigma, \tag{6.5}$$

where $\epsilon_D$ is the bare QD energy and in SI-units $h = \mu_B g B$ with Bohr's magneton $\mu_B$ and the gyromagnetic ratio $g$. The corresponding GFs are derived in Appendix A.3 and given in Eq. ( A.28).

In contrast to the formalism outlined in Section 2.1 we now have several leads connected to a common QD. Therefore, the introduction of a single counting field as in Eq. (2.5) is insufficient. We introduce counting fields $\boldsymbol{\lambda} = (\lambda_1, \cdots, \lambda_N, \lambda_S)$ at each interface between the QD and one of the electrodes via the transformation $\Psi_{\alpha\sigma} \to e^{i\lambda_\alpha(t)/2}\Psi_{\alpha\sigma}$ (Schmidt [ 2007]) so that

$$T_D^{\boldsymbol{\lambda}(t)} = \sum_{\alpha,\sigma} \gamma_\alpha \left[ e^{i\lambda_\alpha(t)/2} d_\sigma^+ \Psi_{\alpha\sigma}(x=0) + \text{H.c.} \right] + \sum_\sigma \gamma_S \left[ e^{i\lambda_S(t)/2} d_\sigma^+ c_\sigma(x=0) + \text{H.c.} \right]. \quad (6.6)$$

Now we can apply again the procedure outlined in Section 3.1 which allows to express the CGF in the form given in Appendix B in Eq. ( B.2).
The first difference compared to the QPCs in Chapters 3 , 4 and 5 is the fact that between the leads and the QD we do not have transparencies but tunnel rates with the unit energy, since the QD represents a single energy level and not a continuum. We define the dot-lead tunnel rates

$$\begin{array}{ll} \Gamma_i = \pi\rho_{0i}\gamma_i^2/2, & \Gamma_{S1} = \pi\rho_{0S}\gamma_S^2/2 \\ \Gamma_S = \Gamma_{S1}|\omega|/(\sqrt{\omega^2-\Delta^2}), & \Gamma_{S2} = \Gamma_{S1}|\Delta|/(\sqrt{\Delta^2-\omega^2}) \end{array}, \quad (6.7)$$

which are affected by the energy-dependent SC DOS.
The interpretation of the result in Eq. (B.2) can be done as for the normal metal-SC QPC in Section 3.1 . For$|\omega| > \Delta$ the counting factors $e^{i(\lambda_i-\lambda_j)}$ describe single-electron transport between the different terminals. For $|\omega| < \Delta$ the SC DOS only allows excitations from the Cooper pair condensate which leads to different transport mechanisms. They are described by counting factors $e^{2i(\lambda_S-\lambda_i)}$ and $e^{2i\lambda_S-i\lambda_i-i\lambda_j}$ referring to the transfer of two particles from the SC to a single or two separate terminals. The transmission coefficients $T_{\text{A}i\sigma}$ (Eq. (B.6)) and $T_{\text{CA}ij\sigma}$ (Eq. B.8 ) therefore refer to AR and CPS, respectively.
For $\Delta \to 0$ the CGF reduces to the well-known expression for the CGF of a non-interacting QD in a multi-terminal geometry (Schmidt et al. [ 2007b]). The same holds for$\gamma_S \to 0$.
The CGF for a three-terminal structure at $T=0$ using a chaotic cavity with energy-independent transmission instead of a resonant level has been calculated before (Börlin et al. [ 2002]). In this case the CGF adopts a characteristic double square root form. The first square root instead of a logarithm is due to the diffusive transport through a chaotic cavity (Belzig and Nazarov [ 2001b]). The second square root may be explained by looking at physical observables that are calculated via derivatives of the CGF, where it leads to additional factors 1/2. Therefore it corresponds to the factors 1/2 in the transparencies $\Gamma_i$ and $\Gamma_{S1}$ that are due to the separate treatment of electrons and holes as a consequence of the proximity effect. The transmission coefficients are different in our case because of the energy-dependent DOS of the QD. As in our case the ones for single-electron transmission, local AR and CPS are proportional to $\Gamma_1\Gamma_2$, $(\Gamma_1^2+\Gamma_2^2)\Gamma_{S2}^2$ and $2\Gamma_1\Gamma_2\Gamma_{S2}^2$, respectively.
Considering the case of only a single normal electrode the conductivity at low voltages and no magnetic field is given by

$$G_{\text{NQS}} = 4\, T_{\text{A}1\uparrow}|_{h=0,\,\omega=0}, \quad (6.8)$$

$$= \frac{4e^2}{h}\left(\frac{2\tilde{\Gamma}_1\tilde{\Gamma}_S}{4\epsilon_D^2+\tilde{\Gamma}_1^2+\tilde{\Gamma}_S^2}\right). \quad (6.9)$$

In the last step we introduced $\tilde{\Gamma}_1 = 2\Gamma_1$, $\tilde{\Gamma}_S = 2\Gamma_{S1}$ and restored SI-units. The result coincides with the one previously obtained in Beenakker [ 1992].
The three-terminal case with two normal metal drains is of particular interest. Positive cross-correlation between the normal electrodes can be used to probe the existence of entanglement (Morten et al. [ 2008]). Along with the noise the cross-correlation can be calculated as a second derivative of the CGF

$$P_{12}^I = -\frac{1}{\tau}\left.\frac{\partial^2 \ln\chi_{\text{SD}}(\boldsymbol{\lambda})}{\partial\lambda_1\partial\lambda_2}\right|_{\boldsymbol{\lambda}=0}. \quad (6.10)$$

Depending on the choice of couplings and voltages different cross correlations may be observed. These depend on the three different types of transport processes between the normal drains that are present: the direct single-electron current proportional to $\Gamma_1\Gamma_2(V_1 - V_2)$, the one due to local ARs proportional to $2\Gamma_{S2}^2(\Gamma_1^2 V_1 + \Gamma_2^2 V_2)$ or the CPS one proportional to $2\Gamma_1\Gamma_2\Gamma_{S2}^2(V_1 + V_2)$. If the SC terminal is weakly coupled to the QD, cross-correlations will either be dominated by single-electron transmission or, for $V_1 \approx V_2$, by AR to a normal drain leading to negative correlations. There are two cases in which positive cross-correlations are observed. If the SC is coupled better to the QD than the normal terminals, positive correlations may be observed for voltages close to the SC gap and $V_1 \approx -V_2$, see Figure 6.2 (a).

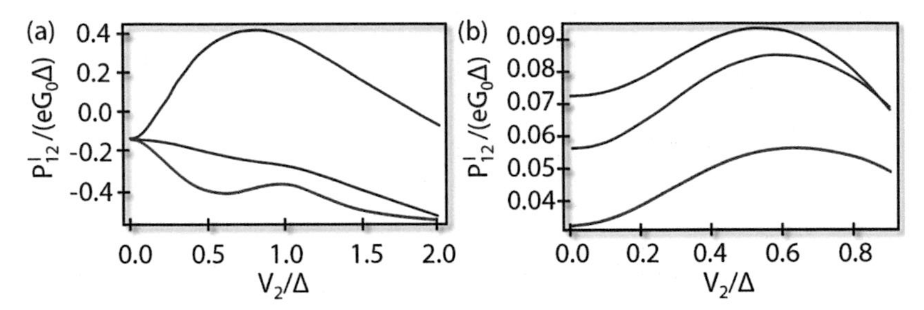


**Figure 6.2.: (a)**: Cross-correlation $P_{12}^I$ calculated from the CGF in equation (B.2) with the parameters $\Gamma_{S1} = \Delta/2 = \Gamma_1$, $\Gamma_2 = 0.2\Delta$, $\epsilon_{D\sigma} = 0$, $T = 0.1\Delta$ as a function of $V_2$ for $V_1 = -V_2$ (blue), $V_1 = 0$ (red) and $V_1 = V_2$ (green). The three curves show that one observes negative cross-correlation for $V_1 = V_2$ and $V_1 = 0$ whereas a positive cross-correlation maximum for $V_1 = -V_2$ close to the SC gap caused by AET is obtained.
**(b)**: Cross-correlation $P_{12}^I$ with the parameters $\Gamma_{S1} = \Delta$, $\Gamma_1 = 0.4\Delta$, $\epsilon_{D\sigma} = 0$, $T = 0.1\Delta$, $V_1 = 0$ as a function of $V_2$. The red curve shows the result for $\Gamma_2 = 0.05\Delta$, the blue curve is for $\Gamma_2 = 0.1\Delta$ and the green curve is for $\Gamma_2 = 0.15\Delta$. Consequently, the second electrode is weakly coupled. Therefore the CPS processes dominate for voltages below the gap over single-electron transmission and AR leading to a positive cross-correlation. This effect is enhanced for weaker coupling of the second electrode. The effect is also present at finite $V_1$ but it is weakened.

In this case CPS is strongly suppressed and one expects single-electron transmission to be dominant. However, the energy-dependent DOS of the SC leads to large transmission coefficients for double ARs from one normal drain to the SC and further to the second normal electrode. This Andreev-reflection enhanced transmission (AET) is also observed if an additional broadening of the SC DOS is taken into account as in Dynes et al. [ 1978], where $\omega \rightarrow \omega + i\Gamma_D$ with a typical experimental value $\Gamma_D$ of about $10^{-2}\Delta$. AET is a robust phenomenon (Freyn et al. [ 2010], Soller and Komnik [ 2011b]). In modern experiments using either InAs nanowires or carbon nanotubes $\Gamma_{S1} \approx \Delta$ is generically obtained (De Franceschi et al. [ 2010]). The hybridisation with the normal leads can be tuned via top gates to e.g. $\Gamma_1 \approx \Gamma_{S1}/2$. AET should then be observable via cross-correlations or from the direct currents.
The second case for positive correlations is observed for strong coupling of the SC to the QD and asymmetrically coupled normal terminals. Since the contribution by CPS is proportional to $V_1 + V_2$ a voltage bias may be applied via the weakly coupled normal electrode $j = 1, 2$. The AR contribution is proportional to $\Gamma_j^2$ and therefore will be strongly suppressed for this weakly coupled drain. For voltages well below the gap CPS dominates over single-electron transmission because a CPS process is possible in two ways. In Figure 6.2 (b) we show the cross-correlation in this situation. The directions of the currents are different in the case of CPS and AET. CAR describes electron

transfer into/from the SC whereas in the case of AET the SC only assists electron transfer between the normal drains. The CGF allows us to follow the different transport processes independently. In the case of CPS two electrons from the same Cooper pair are transferred to spatially separate electrodes. In the case of AET an electron impinging on the SC is retroreflected as a hole. The same Cooper pair that was generated by this AR is then transferred further by a hole from the second normal drain that is retroreflected as an electron. In the case of CPS we therefore observe entangled electrons in the normal electrodes both in energy and spin space. Considering AET the electron and the hole originating from one Cooper pair (due to the conservation of the respective quantities) must also be entangled in energy and spin space. Only AET and CAR generate positive correlations which therefore indicate the presence of entanglement.
The asymmetric coupling necessary to observe CPS can easily be realized experimentally by fabricating a simple SC-QD-normal structure on a substrate with a single normal drain and using a STM tip as the second weakly coupled terminal.
There are two further possibilities to observe CPS as the dominant transport channel. On the one hand using additional normal drains enhances the possibilities for CPS as can be seen directly from the CGF in Eq. (B.2). The combinatorical factor naturally leads to a dominance of CPS. On the other hand AR and single-electron transmission can be fully suppressed by FM drains with antiparallel polarisation (Morten et al. [ 2008]). In this case we always observe positive correlations for voltages below the gap and not too high temperatures if the polarisations are chosen strong enough. For perfect polarization of the FM leads, when $P_1 = -P_2 = 1$ we also considered the case of a normal terminal instead of a SC and obtained $P^I_{12} = 0$ in accordance with previous calculations (Di Lorenzo and Nazarov [ 2005]).

## 6.2. P-wave splitting

Thus far we have considered FMs only via the spin-dependent DOS. However, from Section 3.3 we know about the importance of spin-active scattering at the FM interface. Indeed, spin-active phenomena could be relevant, since up to now we have demonstrated access only to the state $|\Psi^-\rangle$ (see Eq. (6.1)), which is a spin-symmetric combination. There are further maximally entangled states of two spins (Bell states) such as

$$|\Phi^\pm\rangle = 1/\sqrt{2}(|\downarrow\rangle_A|\downarrow\rangle_B \pm |\uparrow\rangle_A|\uparrow\rangle_B), \tag{6.11}$$

involving fully spin-polarized combinations of the two electrons. These $p$-wave components could be generated by spin-active scattering as in Section 4.1 , if we could flip one of the electron spins in Eq. (6.1). Thus, $p$-wave Cooper pair splitters represent an essentlal counterpart to $s$-wave Cooper pair splitters as discussed in Section 6.1 as on-chip sources of spin-entangled EPR pairs.
We first discuss the simplest approach to $p$-wave CPS via a $p$-wave SC and will then discuss how to imitate such a device using spin-active scattering.

### 6.2.1. Superconductor-ferromagnet beamsplitters

Splitting of spin-polarized $p$-wave Cooper pairs can easily be identified in the conductance. From the result in Section 6.1 we find the generalization of Beenakker's formula ( Beenakker [ 1992]) for zero-bias conductance of a beamsplitter realized by a resonant level between a SC and two FMs

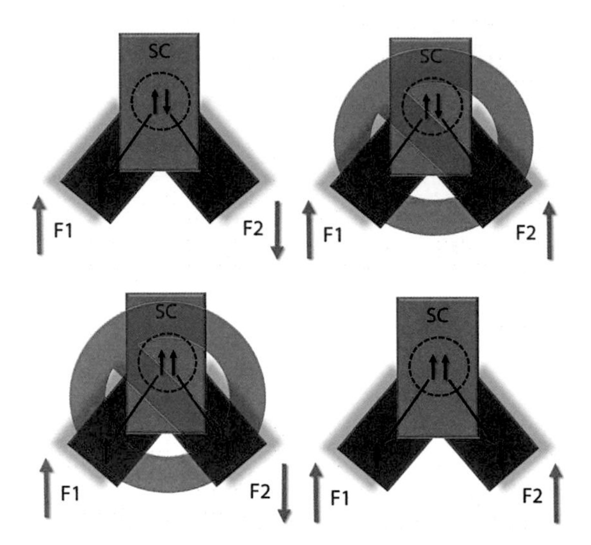


**Figure 6.3.:** Summary of the possible charge transfer processes in a SC-FM beamsplitter. The SC (blue) and the two FMs (red, are assumed to be fully polarized) are coupled via a QD. The polarisation is indicated by green arrows. In the upper part, the situation for $s$-wave SCs is shown. The Cooper pair may split if the two FMs are antiparallely polarized. In the lower part the reversed situation for spin-polarized $p$-wave SCs is depicted. The Cooper pair may now split only if the two FMs are equally polarized.

(without magnetic field, $h = 0$)

$$G_{\text{CPS}}(\sigma_1, \sigma_2) = \frac{8\tilde{\Gamma}_1^2\tilde{\Gamma}_S^2(1+\sigma_1 P_1)(1+\sigma_2 P_2)}{\left[\epsilon_D^2 + \tilde{\Gamma}_S^2 + \tilde{\Gamma}_1^2(2+\sigma_1 P_1+\sigma_2 P_2)^2\right]\left[\epsilon_D^2 + \tilde{\Gamma}_S^2 + \tilde{\Gamma}_1^2(2-\sigma_1 P_1-\sigma_2 P_2)^2\right]}, \tag{6.12}$$

where $\tilde{\Gamma}_1$ is the tunnel rate through the two barriers between the QD and the FM (both are assumed to be equal) as in Eq. (6.9).
In usual $s$-wave superconductivity the spin directions obey $\sigma_1 = -\sigma_2$, and we may maximize CPS by choosing $P_1 = 1 = -P_2$ (or vice versa), see Morten et al. [ 2008]. However, if we choose the polarisations of the FMs to be equal (i.e. $P_1 = 1 = P_2$) $G_{\text{CPS}}$ in Equation (6.12) becomes zero (see Fig. 6.3 ). The situation is reversed if we introduce a spin-polarized $p$-wave SC so that $\sigma_1 = \sigma_2$. Now, splitting is maximized if $P_1 = P_2$. Of course now, for antiparallel polarisation the current is blocked (see Fig. 6.3 ). Therefore $p$-wave splitting is easily identified in the crossed conductance. The assumption of maximal polarisation is not easy to realize in experiment (Löfwander et al. [ 2010]) but we use it to illustrate the argument.
In spite of the apparent simplicity of this consideration a serious problem remains: $p$-wave superconductors are very rare and possible candidates such as $Sr_2RuO_4$ are hard to handle (Murakawa et al. [ 2004]).

### 6.2.2. Cumulant generating function with spin-active scattering

One way to avoid the necessity of $p$-wave SCs is to use setups with spin-active scattering. To identify the separate charge transfer processes the aim is to calculate the CGF. We model the SC-FM beamsplitter by a SC and two FMs $F1$ and $F2$ which are tunnel coupled to a non-interacting QD. The result is the Hamiltonian

$$H_{\text{FBS}} = H_{F1} + H_{F2} + H_{\text{TD1}} + H_{\text{TD2}} + H_S + H_{T1} + H_{T2} + H_D. \tag{6.13}$$

In the following we restrict ourselves to a QD on resonance $\epsilon_D = 0$.
Using the Stoner model in Eq. (3.21) the FM electrodes are described in the language of electron field operators $\Psi_{k\alpha\sigma}$, where $\alpha = F1, F2$. We express the energies of the dot and the reservoirs relative to the SC chemical potential so that $\mu_S = 0$ and $V_\alpha = \mu_S - \mu_\alpha = -\mu_\alpha$ is the chemical potential of the FMs. Tunneling between the SC and the QD is parametrised by $\gamma_S$ as in Eq. (6.4). Finally, we need to introduce spin-active scattering at each of the interfaces between the QD and a FM as in Eq. (3.24)

$$\begin{aligned} H_{\text{TF1},\alpha} &= \sum_\sigma \tilde{\gamma}_{\alpha,1}[\tilde{d}_\sigma^+ \Psi_{\alpha,\sigma}(x=0) + \text{H.c.}], \\ H_{\text{TF2},\alpha} &= \sum_\sigma \tilde{\gamma}_{\alpha,2}[\tilde{d}_\sigma^+ \Psi_{\alpha,-\sigma}(x=0) + \text{H.c.}]. \end{aligned} \tag{6.14}$$

$H_{\text{TF1},\alpha}$ describes normal spin-conserving tunneling, whereas $H_{\text{TF2},\alpha}$ refers to the spin-flip processes [$\Psi_{\alpha-\sigma}(x=0)$ represents an electron in the FM with opposite spin]. If we take spin-active scattering into account this way we have five different tunnel couplings. In order to reduce the number of parameters and for clarification of the discussion we want to limit ourselves to a special constellation of parameters as far as spin-active scattering is concerned, namely $\tilde{\gamma}_{\text{F1,1}} = \tilde{\gamma}_{\text{F2,1}}$, $\tilde{\gamma}_{\text{F1,2}} = \tilde{\gamma}_{\text{F2,2}}$. In this case we define operators

$$d_\sigma = \frac{\tilde{\gamma}_{\text{F1,1}}\tilde{d}_\sigma + \tilde{\gamma}_{\text{F1,2}}\tilde{d}_{-\sigma}}{\sqrt{\tilde{\gamma}_{\text{F1,1}}^2 + \tilde{\gamma}_{\text{F1,2}}^2}}, \tag{6.15}$$

which allow us to rewrite the tunneling Hamiltonians for the SC and (6.14) as in Eq. (6.13) using Eq. (6.3) and

$$\begin{aligned} H_{T1} &= \sum_\sigma \gamma_{S1}[d_\sigma^+ c_\sigma(x=0) + \text{H.c.}], \\ H_{T2} &= \sum_\sigma \gamma_{S2}[d_\sigma^+ c_{-\sigma}(x=0) + \text{H.c.}], \end{aligned} \tag{6.16}$$

with the tunnel amplitudes given by

$$\gamma_{S1} = \frac{\gamma_S \gamma_{\text{F1,1}}}{\sqrt{\gamma_{\text{F1,1}}^2 + \gamma_{\text{F1,2}}^2}}, \quad \gamma_{S2} = \frac{\gamma_S \gamma_{\text{F1,2}}}{\sqrt{\gamma_{\text{F1,1}}^2 + \gamma_{\text{F1,2}}^2}}. \tag{6.17}$$

This minimal model still reveals all transport properties we want to discuss here.
In order to access the CGF, we introduce the counting fields for the leads attached to the QD via the transformations

$$c_\sigma(x=0) \to e^{-i\lambda_S(t)/2} c_\sigma(x=0), \quad \Psi_{\alpha\sigma}(x=0) \to e^{-i\lambda_\alpha(t)/2} \Psi_{\alpha,\sigma}(x=0). \tag{6.18}$$

Following the steps in Section 3.1 the CGF is given by the expectation value

$$\ln \chi_{\mathrm{SFF}}(\boldsymbol{\lambda}) = \left\langle T_C \exp\left[-i \int_C dt \left(T^{\lambda_1}_{\mathrm{TD1}} + T^{\lambda_2}_{\mathrm{TD2}} + T^{\lambda_S}_{T1} + T^{\lambda_S}_{T2}\right)\right]\right\rangle, \tag{6.19}$$

where $\boldsymbol{\lambda} = (\lambda_1, \lambda_2, \lambda_S)$. $T^{\lambda_1}_{\mathrm{TD1}}, \cdots, T^{\lambda_S}_{T2}$ are abbreviations for the tunnelling operators introduced in Equations (6.3) and (6.16) in combination with the substitutions defined in Eq. (6.18).
However, the result for the average in Eq. (6.19) is quite long. We want to restrict ourselves to the study of the relevant aspects of the CGF in view of the possibility of $p$-wave splitting only.

### 6.2.3. Spin-flipped crossed Andreev reflection

In simple SC-FM-QPCs treated in Section 3.3 the presence of spin-active scattering gives rise to a new type of AR. In ordinary AR an electron is retroreflected as a hole with opposite spin since Cooper pairs represent spin singlets. However, due to the spin-active nature of tunneling in the setup considered here the hole or the electron spin can be flipped. SAR induces triplet correlations in the FM as we have seen in Section 3.3. To see whether similar effects occur in our setup we calculate the zero-bias conductance analogously to Equation (6.12) for the case of $P_1 = P_2 = 1$, meaning maximal parallel polarisation

$$G_{\mathrm{SCAR}} = \frac{32\tilde{\Gamma}_1^2\Gamma_{SF}^2(4\Gamma_{\mathrm{SF}}^2 - \Gamma_{\mathrm{SN}}^2)^2}{[\Gamma_{\mathrm{SN}}^4 - 8\Gamma_{\mathrm{SN}}^2\Gamma_{\mathrm{SF}}^2 + 16\Gamma_{\mathrm{SF}}^2(4\tilde{\Gamma}_1^2 + \Gamma_{\mathrm{SF}}^2)]^2}.$$

$\Gamma_{\mathrm{SN}} = \pi\rho_{0S}(\gamma_{S1}^2 + \gamma_{S2}^2)$ and $\tilde{\Gamma}_1 = \pi\rho_0\gamma_\alpha^2$ again refer to the tunnel rates for ordinary single electron tunnelling. $\Gamma_{\mathrm{SF}} = \pi\rho_{0S}\gamma_{S1}\gamma_{S2}$ describes the additional spin-flip tunnel rate at the interface. Mind that $G_{\mathrm{SCAR}} = 0$ for $4\Gamma_{\mathrm{SF}}^2 = \Gamma_{\mathrm{SN}}^2$ is just the result for $V = 0$, $T = 0$ and is leviated for finite bias and/or temperature. Obviously, spin-active scattering has lifted the current blocking indicated in Fig. 6.3 for a $s$-wave SC connected to two maximally parallel polarised ferromagnets. Therefore, this finite conductance is similar to the one obatined for a $p$-wave SC junction in Eq. (6.12). This indicates that this conductance for voltages below the gap is indeed associated to a spin-flipped crossed AR (SCAR) in which a triplet pair is transferred to the ferromagnets.
Of course, one can identify the relevant charge transfer processes directly from the CGF. However, the expression is complicated and the probability distribution of charge transfer is also not easy to access in an experiment as we discussed in Section 2.1. Therefore, we follow the scheme of Section 6.1 and use the cross-correlation to probe whether the two charges of a Cooper pair transferred in a SCAR process are correlated when transferred to the FMs. The presence of non-zero $G_{\mathrm{SCAR}}$ at voltages below the gap and $T = 0$ in combination with a positive cross-correlation can only be explained by a simultaneous transfer of a triplet pair to the ferromagnets, which implies that we indeed observe $p$-wave splitting. The cross-correlation can be calculated as a mixed second derivative of the CGF as in Eq. (6.10)

$$P_{12}^I = -\left.\frac{1}{\tau}\frac{\partial^2 \ln \chi_{\mathrm{SFF}}(\boldsymbol{\lambda})}{\partial\lambda_1\partial\lambda_2}\right|_{\boldsymbol{\lambda}=0}. \tag{6.20}$$

In Fig. 6.4 the result of Equation (6.20) is shown for two different configurations of the couplings and polarisations. For moderate polarisation ($P = 0.3$) and no spin-active scattering we find two cases in which positive cross-correlation can be observed in accordance with our analysis in Section 6.1. First, for voltages close to the SC gap and $V_1 \approx -V_2$ CPS is strongly suppressed and one expects single-electron transmission to be dominant. Nonetheless, the energy-dependent DOS of

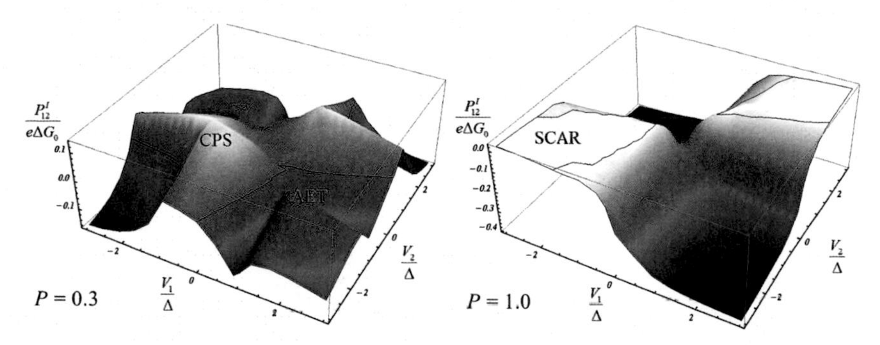


**Figure 6.4.:** Cross-correlation of currents for different parameters: the cross-correlation $P_{12}^I$ according to Equation (6.20) is calculated for two different sets of parameters. The polarisation of the FMs is assumed to be equal in both cases. The left graph shows the result for $P = 0.3$, $\Gamma_{SN} = 4\Delta$, $\Gamma_{SF} = 0$, $\Gamma_1 = 0.4\Delta$, $\Gamma_2 = 0.1\Delta$ and $T = 0.1\Delta$. The right graph is for $P = 1$, $\Gamma_{SN} = 2\Delta$, $\Gamma_{SF} = \Delta$, $\Gamma_1 = 0.05\Delta = \Gamma_2$ and $T = 0.1\Delta$. The cross-correlations are positive in the yellow and white regions.

the SC leads to AET which we identified in Section 6.1 . The second case of positive cross-correlation can be observed for one bias voltage being close to zero and finite bias on the second electrode. In this case, CPS dominates over single-electron transmission and induces a positive cross-correlation as we discussed in Section 6.1 .
This picture changes dramatically if we go over to the case of full polarisation ($P = 1$) and finite spin-active scattering. Since AET relies on double AR, and consequently the spin-symmetry of AR, it must disappear since SAR violates spin-symmetry and it would be the only possible charge transfer process for a single lead for $P = 1$. However, for $V_1 \approx V_2$ a positive cross correlation remains. This is exactly the position, where we assume SCAR to be dominant since $V_1 \approx V_2$ means that single-electron transfer between the FMs does not occur.
The effect is, of course, still observable for $P < 1$ but the polarisation should be rather strong. Spin-active scattering in SC-FM hybrids is a general phenomenon (Hübler et al. [ 2012]) and full polarisation was just assumed for clarity. Therefore, we believe that SCAR is a generic phenomenon that should also be present in SC-FM beamsplitters with tunnel contacts or chaotic cavities since its origin does not lie in the precise form of the energy-dependence of the transmission coefficients. Experiments in the direction of the above described proposal have already been realized. One approach is based on a tunnel coupled SC-FM-SC junction based on Al and Co electrodes with two closely spaced Co-wires bridging two Al-electrodes (Colci et al. [ 2012]). In this experiment the resistance in the case of antiparallel and parallel magnetisation of the two wires has been measured and for low temperatures it has been found that the antiparallel arrangement may even have higher resistance than the parallel one giving reliable evidence for spin-active scattering being present in the device. Concerning multi-terminal hybrid systems with embedded QDs we will discuss a new subgap structure in Chapter 8 that has been observed and can be explained by SAR. This is of special importance, since in our consideration we did not include effects of Coulomb interaction on the dot and so we should worry about a possible suppression of SCAR. The mean field analysis in Chapter 8 , however, will reveal that SAR and therefore also SCAR should be observable also in presence of strong Coulomb interaction. Apart from that, we can argue that also in interacting systems characteristic resonances, as the one of the resonant level considered here, are present and have a characteristic location and width associated with interactions. Therefore, the general scenario

should be robust. Concerning a possible experimental realisation using QDs one should mind that interaction should be small enough and the polarisation large enough to not completely suppress SCAR and the coupling to the leads should be of the order of the SC gap $\Gamma_1$, $\Gamma_{SN}$, $\Gamma_{SF} \approx \Delta$. Nowadays, this coupling is generically obtained in experiments using InAs nanowires or carbon nanotubes as QDs (De Franceschi et al. [ 2010]).

## 6.3. Interacting case

The next logical step is the investigation of interaction effects in SC hybrids. Although the CGFs for quantum dots with a SC lead are already quite interesting we still have a system quadratic in fermion fields. This is the reason for the similarity of the CGF in Eq. (B.2) to the CGF of the tunnel junction in Eq. (3.16). Electron-electron interactions will lead to correlation effects since they allow the electrons to 'see' each other and therefore lead to correlated tunnel processes as we will see below. Interactions play a vital role in reduced dimensions, therefore it seems reasonable to introduce them on the dot only.
From the theoretical side it seems tempting to analyse a limiting case from the onset. Examples include the $U \to \infty$ limit which was analysed in Clerk et al. [ 2000] and Fazio and Raimondi [1998] and the $\Delta \to \infty$ limit which was analysed via a rate equation approach in Braggio et al. [2011]. These mappings of the problem to specific, again non-interacting, cases cannot show specific correlation effects. Furthermore, in actual experiments these limits can never be completely justified since typically the tunnel rates are of the order of $\Delta$ (De Franceschi et al. [ 2010]) and$U/\Delta$ can be around 6.5 (Cuevas et al. [ 2001]). We will analyse the$U \to \infty$ limit in more detail in Section 6.4 and show that despite the abovementioned shortcomings its results allow for reliable predictions for experiments. The rate equation approach will be reviewed in more detail in Chapter 7 .
First, however, we will follow the, so far, only approach to the full complexity of the Anderson model with a SC lead presented in Cuevas et al. [ 2001] and Yamada et al. [ 2011]. In these works the authors use an interpolative scheme for the electron self-energy, which has a similar form in the weak and strong interaction case. For the case of weak interaction they perform perturbation theory to second order in the Coulomb interaction. Since we aim at calculating the CGF this calculation represents the first step (Gogolin and Komnik [ 2006b]). A numerical evaluation of the CGF is sufficient for calculating the relevant cumulants. In order to identify the relevant charge transfer processes an analytical calculation is needed, which makes it necessary to analyse limiting cases in which analytical progress is possible. The first simplification is to include only one normal lead in order to have a simpler two-terminal geometry. As in the case of the Anderson model with normal leads we take the tunnel rates $\Gamma_{S1} = \Gamma_1$ to be symmetric and the largest energy scale and $T = 0$. Concerning the ratio of the bias $V$ and $\Delta$ we take two limits: either $V \gg \Delta$ or $V \ll \Delta$ corresponding to charge transfer above the SC gap or below. This assumption allows for additional simplification since we can use the approximate GFs for the SC introduced in Eqs. (3.14) and (3.15). Using Eq. (3.10) we obtain the exact-in-tunneling normal and anomalous dot GFs $\mathcal{D}^{\lambda}_{0\sigma}(\omega)$ and $\mathcal{C}^{\lambda}_{0}(\omega), \mathcal{C}^{\lambda+}_{0}(\omega)$ which are given in Appendix B Eqs. ( B.11) and (B.12).
From here, there are two ways to proceed: one can either use a linked cluster expansion (Schmidt et al. [ 2007a]) or directly modify the Dyson equation (Gogolin and Komnik [ 2006b]). We will illustrate both approaches in the following.

### 6.3.1. Voltages above the gap

We will first use the linked cluster expansion for the case $V \gg \Delta$. In this case the calculation can be significantly simplified since for the anomalous GF we have $\mathcal{C}_0^{\lambda+}(\omega) = 0$. Therefore the results will be comparable to the normal-conducting case.
We derived Eq. (2.6) for the CGF under the assumption that the tunneling operator is the perturbation of the Hamiltonian. In order to model the onsite interaction we add to the Hamiltonian in Eq. (6.2) the term

$$H_U = U d_\uparrow^+ d_\uparrow d_\downarrow^+ d_\downarrow = U n_{D\uparrow} n_{D\downarrow}. \tag{6.21}$$

Considering the case $V \gg \Delta$ we express the exact CGF $\ln \chi_{\text{NS,exact,1}}$ as a correction to the non-interacting CGF in Eq. (B.2). The quantity we evaluate is

$$\chi_{\text{NS,exact,1}} = \chi_{\text{SD}}(\lambda) \left\langle T_{\mathcal{C}} \exp\left(-i \int_{\mathcal{C}} dt\ H_U(t)\right)\right\rangle, \tag{6.22}$$

where $\chi_{SD}$ refers to the two-terminal case of Eq. (B.2) with $\lambda = \lambda_1 - \lambda_S$ and where the expectation value is taken with respect to the non-interacting ground state. One can show via the linked cluster expansion (Schmidt [2007])

$$\chi_{\text{NS,exact,1}} = \chi_{\text{SD}}(\lambda) \exp\left\{\sum_{n=1}^{\infty} \xi_n\right\}, \tag{6.23}$$

where $\xi_n$ contains the connected diagrams with $n$ vertices

$$\xi_n = \frac{(-i)^n}{n!} \int_{\mathcal{C}} dt_1 \cdots dt_n \langle T_{\mathcal{C}} H_U(t_1) \cdots H_U(t_n)\rangle_\lambda. \tag{6.24}$$

This means that the CGF allows for a series expansion

$$\ln \chi_{\text{NS,exact,1}} = \ln \chi_{SD}(\lambda) + \sum_{n=1}^{\infty} \ln \chi_1^{(n)}(\lambda). \tag{6.25}$$

The first order in $U$ is given by

$$\ln \chi_1^{(1)}(\lambda) = -i \int_{\mathcal{C}} dt \langle T_{\mathcal{C}}[d_\uparrow^+(t) d_\uparrow(t) - n_0]\rangle_\lambda \langle T_{\mathcal{C}}[d_\downarrow^+(t) d_\downarrow(t) - n_0]\rangle_\lambda. \tag{6.26}$$

The expression factorises with respect to spin to

$$\ln \chi_1^{(1)}(\lambda) = -iU \int_{\mathcal{C}} dt\ N_\uparrow(t) N_\downarrow(t), \tag{6.27}$$

where we have used the normalised particle density on the dot

$$N_\sigma(t) := \langle T_{\mathcal{C}} d_\sigma^+(t) d_\sigma(t)\rangle_\lambda - n_0. \tag{6.28}$$

The symmetric Anderson model corresponds to $n_0 = 1/2$ and $\epsilon_{D\sigma} = V/2$ for $h = 0$ (Yosida and Yamada [1975]). Expressing $\langle T_{\mathcal{C}} d_\sigma^+(t) d_\sigma(t)\rangle_\lambda$ using $\mathcal{D}_0^\lambda(\omega)$ we arrive at

$$\ln \chi_1^{(1)}(\lambda) \quad = \quad -\frac{U \tau V}{\pi^2 \Gamma_1^3} \sum_\sigma (\epsilon_{D\sigma} - V/2)(\epsilon_{D\bar{\sigma}} - V/2)(e^{-i\lambda} - 1). \tag{6.29}$$

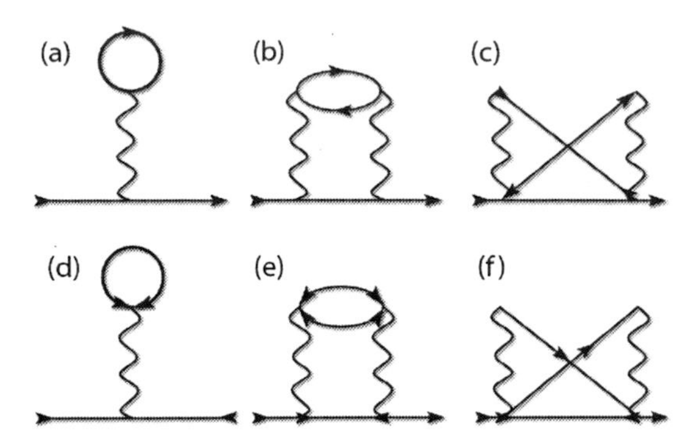


**Figure 6.5.:** Modified dot GFs in the superconducting state: lines denote either normal or anomalous exact-in-tunneling GFs and the wiggly lines correspond to onsite interactions.

The first order result is therefore an effect of broken particle-hole symmetry either by an applied magnetic field or by a shifted dot level energy. Although the first order adds to the statistics it does not produce any novel transport processes.
The corresponding modified GF is shown in Fig. 6.5 (a). The only diagram to second order which is not due to broken particle-hole symmetry is shown in Fig. 6.5 (b).
The presence of such a diagram points to the fact that genuine interaction induced processes can be expected to second order in $U$. Also for the second order we start from the linked cluster expansion

$$\ln \chi_1^{(2)}(\lambda) = -\frac{U^2}{2} \int_C dt\, dt' \, \langle T_C \, d_\uparrow^+(t) d_\uparrow(t) d_\uparrow^+(t') d_\uparrow(t') d_\downarrow^+(t) d_\downarrow(t) d_\downarrow^+(t') d_\downarrow(t') \rangle_\lambda . \tag{6.30}$$

Calculating all connected diagrams emerging from Eq. (6.30) we arrive at three terms

$$\begin{aligned} \ln \chi_1^{(2)}(\lambda) &= -\frac{U^2}{2} \int_C dt\, dt' \left[ \sum_\sigma N_\sigma(t) N_\sigma(t') \mathcal{D}_{0\bar\sigma}^\lambda(t-t') \mathcal{D}_{0\bar\sigma}^\lambda(t'-t) \right. \\ &\quad \left. + \mathcal{D}_{0\uparrow}^\lambda(t-t') \mathcal{D}_{0\uparrow}^\lambda(t'-t) \mathcal{D}_{0\downarrow}^\lambda(t-t') \mathcal{D}_{0\downarrow}^\lambda(t'-t) \right], \end{aligned} \tag{6.31}$$

where the latter term indeed couples the two spins and corresponds to the modified dot GF in Fig. 6.5 (b). Proceeding as in Schmidt [2007] and adding all contributions which arise to second order in $U$ we can conclude that up to order $O(\Gamma^{-4})$ the correction to the CGF for $V \gg \Delta$ is given by

$$\begin{aligned} \ln \chi_1^{(2)}(\lambda) &= \frac{U^2 \tau V}{2\pi^3 \Gamma_1^4} \sum_\sigma \left( \epsilon_{D\bar\sigma} - \frac{V}{2} \right)^2 (e^{-i\lambda} - 1) \\ &\quad + \frac{U^2 \tau V}{\pi^3 \Gamma_1^4} \left( 3 - \frac{\pi^2}{4} \right) \sum_\sigma \left( \epsilon_{D\sigma} - \frac{V}{2} \right)^2 (e^{-i\lambda} - 1) \\ &\quad + \frac{U^2 \tau V^3}{24 \pi^3 \Gamma_1^4} \left[ 4(e^{-2i\lambda} - 1) + \sum_\sigma (e^{-i\lambda} - 1) \right] \\ &\quad + \frac{U^2 \tau V^3}{12 \pi^3 \Gamma^4} \left( 3 - \frac{\pi^2}{4} \right) \sum_\sigma (e^{-i\lambda} - 1). \end{aligned} \tag{6.32}$$

The terms in the first two lines are linear in the applied voltage and depend on the magnetic field. They can be interpreted again as effects of broken particle-hole symmetry, similar to the first order in Eq. (6.29). The third line is the most remarkable since it can be interpreted as correlated tunneling of electron pairs which consist of two electrons with opposite spin. This interpretation can be substantiated by considering the spin-dependent case (Schmidt et al. [2007a]) or the Toulouse point of the corresponding Kondo Hamiltonian (Gogolin and Komnik [2006b]). Consequently, the onsite interaction leads to similar correlations between the electrons as superconductivity (see Eq. 2.8). Indeed, these correlations can be identified similar to the ones in SC beamsplitters (Schmidt et al. [2007b]). The effect can be rationalized as two electrons hopping onto the dot, feeling their Coulomb repulsion and being back-reflected coherently.

### 6.3.2. Voltages below the gap

Due to this electron-pairing effect it is necessary to inspect the range of voltages below the gap. In this case we use the modification of the Dyson equation as in Gogolin and Komnik [2006b]. From now on we consider $h = 0$ so that we can drop the spin index. Below the gap the anomalous GF is also present so that the self-energy has two parts (the 1 and 2 component in Nambu space). The full dot GF in Keldysh space is given by

$$\mathcal{D}^\lambda(\omega) = \mathcal{D}_0^\lambda(\omega) + \mathcal{D}^\lambda(\omega)\Sigma_1(\omega)\mathcal{D}_0^\lambda(\omega) + \mathcal{D}^\lambda(\omega)\Sigma_2(\omega)\mathcal{C}_0^{\lambda+}(\omega), \tag{6.33}$$

where $\Sigma_1$ is the exact self-energy caused by modifications to the dot GF as in Fig. 6.5 (a), (b) and (c) and $\Sigma_2$ is caused by dot GFs such as in Fig. 6.5 (d), (e) and (f). We first consider only the contribution by $\Sigma_1$. Neglecting $\Sigma_2$ we can again perform the inversion of $\mathcal{D}^\lambda(\omega)$ and insert it into Eq. (3.7). The correction to the adiabatic potential $\delta_1 U_a$ to second order in $U$ is given by

$$\frac{\partial(\delta_1 U_a)}{\partial\lambda} = -i\Gamma_1 \int_{-\infty}^{\infty} \frac{d\omega}{2\pi \mathrm{Det}(\omega)} \left[e^{i\lambda/2} n_1 \Sigma_1^{+-}(\omega) + e^{-i\lambda/2}(1-n_1)\Sigma_1^{-+}(\omega)\right] \theta\left(\frac{\Delta - |\omega|}{\Delta}\right), \tag{6.34}$$

where $\mathrm{Det}(\omega)$ refers to the determinant of $\mathcal{D}_0^\lambda(\omega)$. We calculate the lowest order contributions to the self-energy (Fig. 6.5 (a), (b) and (c)) in the time domain

$$\begin{aligned}\Sigma_1(t) = \Bigg[ & -iU\delta(t)\mathcal{D}_0^{\lambda--}(0) + U^2[\mathcal{D}_0^{\lambda--}(t)]^2\mathcal{D}_0^{\lambda--}(-t) + U^2\mathcal{C}_0^{\lambda--}(t)\mathcal{C}_0^{\lambda+--}(t)\mathcal{D}_0^{\lambda--}(-t) \\ & -U^2[\mathcal{D}_0^{\lambda+-}(t)]^2\mathcal{D}_0^{\lambda-+}(-t) - U^2\mathcal{C}_0^{\lambda++-}(t)\mathcal{C}_0^{\lambda++-}(t)\mathcal{D}_0^{\lambda-+}(-t) \\ & -U^2[\mathcal{D}_0^{\lambda-+}(t)]^2\mathcal{D}_0^{\lambda+-}(-t) - U^2\mathcal{C}_0^{\lambda-+}(t)\mathcal{C}_0^{\lambda+-+}(t)\mathcal{D}_0^{\lambda+-}(-t) \\ & iU\delta(t)\mathcal{D}_0^{\lambda++}(0) + U^2[\mathcal{D}_0^{\lambda++}(t)]^2\mathcal{D}_0^{\lambda++}(-t) \Bigg].\end{aligned} \tag{6.35}$$

The linear part in $U$ is a remnant of the occupation probability of the dot level and can be evaluated as

$$-iU\mathcal{D}_0^{\lambda--}(0) = -iU\mathcal{D}_0^{\lambda-+}(0) = -iU\int_{-\Delta}^{\Delta}\frac{d\omega}{2\pi}\mathcal{D}_0^{\lambda-+}(\omega) = Un_\lambda. \tag{6.36}$$

Using this result for $n_\lambda$ in Eq. (6.34) leads to an expression for the total CGF which is identical to Eq. (B.2) up to the result for the denominator where the bare level-energy is renormalised by $\tilde{\epsilon}_{D\sigma} \to \epsilon_{D\sigma} + Un_\lambda$. Subsequent expansion in $U$ and integration over energy results in a well controlled contribution which vanishes for the symmetric Anderson model $\epsilon_{D\sigma} = -U/2$ (Wiethege et al. [1982]) to which case the following considerations are restricted.

We now concentrate on the contribution to second order in $U$. As in (Yamada et al. [2011]) we proceed by the calculation of the necessary susceptibilities for the self-energy

$$\chi_{01}(\Omega) = i\int_{-\infty}^{\infty}\frac{d\omega}{2\pi}\begin{bmatrix} \mathcal{D}_0^{\lambda--}(\omega+\Omega)\mathcal{D}_0^{\lambda--}(\omega) & \mathcal{D}_0^{\lambda-+}(\omega+\Omega)\mathcal{D}_0^{\lambda+-}(\omega) \\ \mathcal{D}_0^{\lambda+-}(\omega+\Omega)\mathcal{D}_0^{\lambda-+}(\omega) & \mathcal{D}_0^{\lambda++}(\omega+\Omega)\mathcal{D}_0^{\lambda++}(\omega)\end{bmatrix}, \tag{6.37}$$

$$\chi_{02}(\Omega) = i\int_{-\infty}^{\infty}\frac{d\omega}{2\pi}\begin{bmatrix} \mathcal{C}_0^{\lambda+--}(\omega+\Omega)\mathcal{D}_0^{\lambda--}(\omega) & \mathcal{C}_0^{\lambda+-+}(\omega+\Omega)\mathcal{D}_0^{\lambda+-}(\omega) \\ \mathcal{C}_0^{\lambda++-}(\omega+\Omega)\mathcal{D}_0^{\lambda-+}(\omega) & \mathcal{C}_0^{\lambda+++}(\omega+\Omega)\mathcal{D}_0^{\lambda++}(\omega)\end{bmatrix}. \tag{6.38}$$

The respective self-energy can be extracted from

$$\Sigma_1(\omega) = i\int_{-\infty}^{\infty}\frac{d\Omega}{2\pi}\begin{bmatrix} \mathcal{D}_0^{\lambda--}(\omega-\Omega)\chi_{01}^{--}(\Omega)+\mathcal{C}_0^{\lambda--}(\omega-\Omega)\chi_{02}^{--}(\Omega) & \mathcal{D}_0^{\lambda-+}(\omega-\Omega)\chi_{01}^{-+}(\Omega)+\mathcal{C}_0^{\lambda-+}(\omega-\Omega)\chi_{02}^{-+}(\Omega) \\ \mathcal{D}_0^{\lambda+-}(\omega-\Omega)\chi_{01}^{+-}(\Omega)+\mathcal{C}_0^{\lambda+-}(\omega-\Omega)\chi_{02}^{+-}(\Omega) & \mathcal{D}_0^{\lambda++}(\omega-\Omega)\chi_{01}^{++}(\Omega)+\mathcal{C}_0^{\lambda++}(\omega-\Omega)\chi_{02}^{++}(\Omega)\end{bmatrix}. \tag{6.39}$$

From this expression we calculate $\Sigma_1^{-+}(\omega)$ and $\Sigma_1^{+-}(\omega)$. Integration of Eq. (6.34) with respect to $\lambda$ directly gives one part for the correction to the CGF. The prefactors are given as numbers and all prefactors have been evaluated to leading order in $V$

$$\begin{aligned}\ln\chi_{2,1}^{(2)}(\lambda) = & -0.027\frac{\tau U^2 V\Delta^2}{\Gamma^4}(e^{i\lambda}-1)+0.0016\frac{\tau U^2 V\Delta^2}{\Gamma^4}(e^{-i\lambda}-1) \\ & -0.0053\frac{\tau U^2 V\Delta^2}{\Gamma^4}(e^{-2i\lambda}-1)+\frac{\tau U^2 V\Delta^2}{\Gamma^4}0.007(e^{-3i\lambda}-1) \\ & +\frac{\tau U^2 V^3}{4\pi^2\Gamma^4}(e^{-4i\lambda}-1).\end{aligned} \tag{6.40}$$

Several comments on this result are in order: first, this only represents one part of the correction since so far we have not considered $\Sigma_2$. Second, we indeed observe correlation effects leading to a nontrivial dependence on $\lambda$. The Coulomb repulsion adds the possibility of breaking a Cooper pair on the dot which leads to terms $\propto e^{i\lambda}$, $e^{-i\lambda}$. Additionally such processes may involve an AR leading to terms $\propto e^{-3i\lambda}$. The most interesting term is the one $\propto e^{-4i\lambda}$. In Section 6.3.1 we have interpreted the occurence of double exponential terms as correlated tunneling of electron pairs. In this case correlated tunneling of electron pairs to the SC is prohibited since charge transfer has to proceed via AR so that the correlated tunnel processes result in quartets in the CGF. This interpretation is substantiated by the fact that the final term in Eq. (6.40) is due to the purely normal part of the self-energy and therefore has the same origin as the correlated tunneling terms in Eq. (6.32). The process leading to the quartic contribution is illustrated in Fig. 6.6.
However, such an interpretation is problematic without evaluating the contribution from $\Sigma_2$. In this case we use Eq. (6.33) and use the second term directly in Eq. (3.6) to arrive at

$$\begin{aligned}\frac{\partial(\delta_2 U_a)}{\partial\lambda} = & -i\Gamma_1\int_{-\infty}^{\infty}\frac{d\omega}{2\pi}\Big[e^{i\lambda/2}n_1[\mathcal{D}_0^{\lambda}(\omega)\Sigma_2(\omega)\mathcal{C}_0^{\lambda+}(\omega)]^{+-} \\ & +e^{-i\lambda/2}(1-n_1)[\mathcal{D}_0^{\lambda}(\omega)\Sigma_2(\omega)\mathcal{C}_0^{\lambda+}(\omega)]^{-+}\Big]\,\theta\left(\frac{\Delta-|\omega|}{\Delta}\right).\end{aligned} \tag{6.41}$$

The second self-energy in the time domain is given by

$$\Sigma_2(t) = \begin{bmatrix} -iU\delta(t)\mathcal{C}_0^{\lambda+--}(0)+U^2[\mathcal{C}_0^{\lambda--}(t)\mathcal{C}_0^{\lambda+--}(t)]\mathcal{C}_0^{\lambda--}(-t)+U^2[\mathcal{D}_0^{\lambda--}(t)\mathcal{C}_0^{\lambda--}(t)]\mathcal{D}_0^{\lambda--}(-t) & -U^2[\mathcal{D}_0^{\lambda-+}(t)\mathcal{C}_0^{\lambda-+}(t)]\mathcal{D}_0^{\lambda+-}(-t)-U^2[\mathcal{C}_0^{\lambda-+}(t)\mathcal{C}_0^{\lambda+-+}(t)]\mathcal{C}_0^{\lambda+-}(-t) \\ -U^2[\mathcal{D}_0^{\lambda+-}(t)\mathcal{C}_0^{\lambda++-}(t)]\mathcal{C}_0^{\lambda-+}(-t)-U^2[\mathcal{C}_0^{\lambda+-}(t)\mathcal{C}_0^{\lambda++-}(t)]\mathcal{D}_0^{\lambda-+}(-t) & iU\delta(t)\mathcal{C}_0^{\lambda+++}(t)+U^2[\mathcal{C}_0^{\lambda++}(t)\mathcal{C}_0^{\lambda+++}(t)]\mathcal{C}_0^{\lambda++}(-t)+U^2[\mathcal{D}_0^{\lambda++}(t)\mathcal{C}_0^{\lambda++}(t)]\mathcal{D}_0^{\lambda++}(-t)\end{bmatrix}.$$

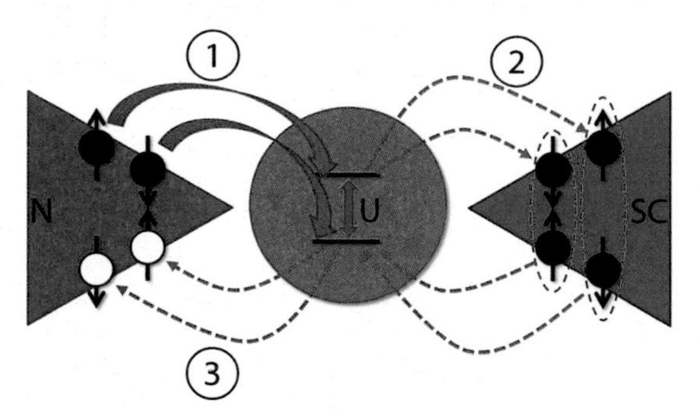


**Figure 6.6.:** Illustration of the quartic contribution to the CGF: two electrons hop onto the quantum dot (1) and 'feel' their mutual repulsion $U$. This repulsion leads to correlated scattering onto the SC which results in two correlated ARs (2). The two holes are transferred further to the normal lead (3) which leads to a quartic contribution to the CGF.

The part linear in $U$ is a remnant of the finite SC gap on the dot $\langle d_\uparrow^+ d_\downarrow^+ \rangle$. As for the normal GF we can evaluate it via

$$-iU\mathcal{C}_0^{\lambda+--}(0) = -iU\int\frac{d\omega}{2\pi}\mathcal{C}_0^{\lambda+--}(\omega). \tag{6.42}$$

The contribution we obtain is always finite and does not vanish for the symmetric Anderson model. However, insertion of this result into Eq. (6.41) and integration with respect to $\lambda$ leads to a well controlled contribution which is of the same form as the original CGF in Eq. (B.2). Therefore, this contribution can be taken into account as an effective change of the local pairing potential $\Delta \to \Delta_{\text{eff}}$ as it has been done in previous works (Vecino et al. [ 2003]).
Next, we evaluate the second order contributions in $U$. We need to evaluate one more susceptibility

$$\chi_{03}(\Omega) = i\int_{-\infty}^{\infty}\frac{d\omega}{2\pi}\begin{bmatrix} \mathcal{C}_0^{\lambda+--}(\omega+\Omega)\mathcal{C}_0^{\lambda--}(\omega) & \mathcal{C}_0^{\lambda+-+}(\omega+\Omega)\mathcal{C}_0^{\lambda+-}(\omega) \\ \mathcal{C}_0^{\lambda++-}(\omega+\Omega)\mathcal{C}_0^{\lambda-+}(\omega) & \mathcal{C}_0^{\lambda+++}(\omega+\Omega)\mathcal{C}_0^{\lambda++}(\omega) \end{bmatrix}. \tag{6.43}$$

The second part of the self-energy can then be extracted from

$$\begin{aligned}\Sigma_2(\omega) = i\int_{-\infty}^{\infty}\frac{d\Omega}{2\pi}&\left[\begin{matrix} \mathcal{D}_0^{\lambda--}(\omega-\Omega)\chi_{02}^{--}(\Omega)+\mathcal{C}_0^{\lambda--}(\omega-\Omega)\chi_{03}^{--}(\Omega) \\ \mathcal{D}_0^{\lambda+-}(\omega-\Omega)\chi_{02}^{+-}(\Omega)+\mathcal{C}_0^{\lambda+-}(\omega-\Omega)\chi_{03}^{+-}(\Omega) \end{matrix}\right.\\ &\left.\begin{matrix} \mathcal{D}_0^{\lambda-+}(\omega-\Omega)\chi_{02}^{-+}(\Omega)+\mathcal{C}_0^{\lambda-+}(\omega-\Omega)\chi_{03}^{-+}(\Omega) \\ \mathcal{D}_0^{\lambda++}(\omega-\Omega)\chi_{02}^{++}(\Omega)+\mathcal{C}_0^{\lambda++}(\omega-\Omega)\chi_{03}^{++}(\Omega) \end{matrix}\right].\end{aligned} \tag{6.44}$$

These self-energies can be put into Eq. (6.41) and integration with respect to $\lambda$ leads to the second contribution to the CGF. In total we obtain the following correction to second order in $U$ for the

CGF

$$\begin{aligned}
\ln \chi_2^{(2)} &= \frac{\tau U^2 V \Delta^2}{\pi^2 \Gamma_1^4}(-0.001015)(e^{3i\lambda}-1) + \frac{\tau U^2 V \Delta^2}{\pi^2 \Gamma_1^4}(0.017975)(e^{2i\lambda}-1) \\
&+ \frac{\tau U^2 V \Delta^2}{\pi^2 \Gamma_1^4}(-0.002713)(e^{i\lambda}-1) + \frac{\tau U^2 V \Delta^2}{\pi^2 \Gamma_1^4}(-0.00307)(e^{-i\lambda}-1) \\
&+ \frac{\tau U^2 V \Delta^2}{\pi^2 \Gamma_1^4}(-0.00307)(e^{-2i\lambda}-1) + \frac{\tau U^2 V \Delta^2}{\pi^2 \Gamma_1^4}(0.0137)(e^{-3i\lambda}-1) \\
&+ \frac{\tau U^2 V \Delta^2}{\pi^2 \Gamma_1^4}(-0.0179)(e^{-4i\lambda}-1). \qquad (6.45)
\end{aligned}$$

We observe, that the quartic contribution does not vanish and even becomes first order in $V$. We may therefore conclude that the production of quartets is straightforward in a simple SC-QD-normal junction due to the onsite interaction and does not require the more sophisticated triple SC junction investigated in Freyn et al. [ 2011]. Indeed, the quartic correlations should be directly observable via the fourth cumulant in a SC-QD junction with four normal leads in direct analogy to (Schmidt et al. [ 2007b]).
The comparisons of our results to others are limited due to the fact that all approaches to second order presented so far (Cuevas et al. [ 2001], Yamada et al. [ 2011]) have been done numerically to extend the range of validity to a bigger voltage range. However, Eq. (6.45) predicts a conductance reduction at zero bias which is of the correct order of magnitude also for $U \approx 2\Gamma_1$. Extending the above result to all orders in $U$ and $\Gamma_1$ has been possible in the normal-conducting case (Gogolin and Komnik [ 2006b]) by using equilibrium Bethe ansatz calculations. However, similar results are not available for the case investigated here.

## 6.4. Kondo case

In view of these difficulties it seems more rewarding to use effective descriptions for a QD in the presence of onsite interaction in order to make contact with contemporary experiments. The most prominent effect of interactions is the Kondo resonance that we have discussed in Section 2.4 . It is observed when the number of electrons on the QD is odd so that it obtains a localized magnetic moment of spin 1/2. At energy-scales below the Kondo temperature $T_K$ the dot spin hybridizes with the lead spin density and forms a singlet state. At this strong-coupling fixed point a perfectly transmitting channel opens up and the Kondo effect can be described as a pure resonant level system as far as the electronic transport is concerned. For weak coupling of the second lead to the QD one can see from our investigation of the non-interacting case in Section 6.1 that a resonant level leads to a renormalization of the effective transmission coefficient for a tunnel contact with the DOS of a resonant level. This situation emerges in the case of a QD in the Kondo regime contacted by a normal lead and a STM tip (KNQN) or by a SC provided $T_K \lesssim \Delta$ holds (Gräber et al. [ 2004]) (KNQS). In the opposite case$T_K > \Delta$ the Kondo resonance also strongly couples to the quasiparticles on the SC side, leaving the $I - V$ characteristics almost identical to those obtained for normal conducting systems (Gräber et al. [ 2004]). If only one drain strongly couples to the QD the coupling to the second lead can be described by an effective tunneling transmission. We can thus employ the transmission coefficients for the QPC multiplied with the Kondo DOS $\Gamma_K^2 / \left[(\omega - V)^2 + \Gamma_K^2\right]$, where $\Gamma_K \approx T_K$ (Gräber et al. [ 2004]). The investigation of such an effective model for the Kondo situation as far as the CGF is concerned has been carried out in (Soller and Komnik [ 2011a]).

If the Kondo impurity is contacted by two normal drains with strongly asymmetric coupling its transmission can be effectively written as

$$T_{\mathrm{enK}} = \frac{4\Gamma}{(1+\Gamma)^2} \frac{\Gamma_K^2}{\left[(\omega - V)^2 + \Gamma_K^2\right]}, \tag{6.46}$$

where $T_{\mathrm{en}} = 4\Gamma/(1+\Gamma)^2$ is the transmission coefficient of the ordinary QPC. The factor $\Gamma_K^2/[(\omega - V)^2 + \Gamma_K^2]$ accounts for the Kondo resonance, which is pinned to the Fermi level of the strongly coupled drain.
To describe actual experiments we have to include the DOS outside the Kondo peak given e.g. by the Hubbard subbands at $\pm U/2$. In a first approximation this can be modelled by an energy-independent transmission coefficient $T_g$ as in Gräber et al. [ 2004]. The FCS of these additional processes can be described by the standard Levitov-Lesovik formula (Levitov and Lesovik [ 1994])

$$\ln \chi_g(\lambda, \tau) = 2\tau \int \frac{d\omega}{2\pi} \ln\{1 + T_g[(e^{i\lambda} - 1)n_{1+}(1 - n_S) + (e^{-i\lambda} - 1)n_S(1 - n_{1+})]\}. \tag{6.47}$$

The total CGF is given by $\ln \chi_{\mathrm{KN}} = \ln \chi_{\mathrm{KNQN}} + \ln \chi_g$, where $\chi_{\mathrm{KNQN}}$ can be derived from $\chi_g$ by replacing $T_g$ by $T_{\mathrm{enK}}$.
For the differential conductance we observe the typical Lorentzian shape that is given by the form of the Kondo resonance in Eq. (6.46). However, in this case there is no energy scale apart from the width of the Kondo resonance.
This is different in the case of a SC lead with its characteristic energy gap. E.g. for two SC terminals the Kondo effect is suppressed if the SC gap is larger than $T_K$ (Buitelaar et al. [ 2002]). A similar effect occurs in the regime $\Gamma_K \lesssim \Delta$ considered here in a KNQS junction since only electrons on the normal side strongly couple to the emerging Kondo resonance. One therefore obtains a strong asymmetry of the Kondo coupling between the normal and the SC drain to the dot (Gräber et al. [2004]). Consequently, one observes a strong suppression of the total conductance while side peaks at the SC gap appear. The assembly of SC-QD hybrid structures with good coupling to the QD makes $\Gamma_K \lesssim \Delta$ a typical experimental case (De Franceschi et al. [ 2010]).
Since the coupling of the SC to the Kondo impurity is small and branch-crossing and AR above the gap are both processes of higher order we find that we may safely neglect their contributions. We can thus employ the transmission coefficients for the normal-SC QPC in Eq. (3.12) with the Kondo DOS $\Gamma_K^2/\left[(\omega - V)^2 + \Gamma_K^2\right]$ for single-electron transmission and for the ARs we use $\Gamma_K^4/\left\{\left[(\omega - V)^2 + \Gamma_K^2\right]\left[(\omega + V)^2 + \Gamma_K^2\right]\right\}$ to obtain

$$\begin{aligned} T_{\mathrm{eK}}(\omega) &= \frac{4\Gamma_e}{(1+\Gamma_e)^2} \frac{\Gamma_K^2}{(\omega - V)^2 + \Gamma_K^2}, \\ T_{\mathrm{AK}}(\omega) &= \frac{4\Gamma_A^2}{\Gamma_A^4 + 2\Gamma_A^2(1 - \Gamma_e^2) + (1 + \Gamma_e^2)^2} \frac{\Gamma_K^4}{\left[(\omega - V)^2 + \Gamma_K^2\right]\left[(\omega + V)^2 + \Gamma_K^2\right]}. \end{aligned} \tag{6.48}$$

The total CGF, including the background DOS, is given by $\ln \chi_{\mathrm{KS}} = \ln \chi_{\mathrm{KNQS}} + \ln \chi_g$, where $\chi_{\mathrm{KNQS}}$ can be obtained from the CGF in Eq. (3.16) by replacing $T_{\mathrm{e,a}}$ by $T_{\mathrm{eK}}$ and $T_{\mathrm{A,a}}$ by $T_{\mathrm{AK}}$.
We calculate the ratio of the differential conductances in the case of a KNQS junction and a KNQN device. Fig. 6.7 (a) shows the comparison of the theory outlined above with experimental data, where the Kondo conductance peak in the KNQN state has been approximated by a Lorentzian (Gräber et al. [ 2004]). We observe a fairly good agreement up to a small difference for small voltages. This can be attributed to additional diffusive transport channels.
In contrast to the prediction by the semiconductor model (Tinkham [ 1996]) we obtain quantitative agreement for voltages up to approximately $2\Delta$ due to the explicit inclusion of AR and using the

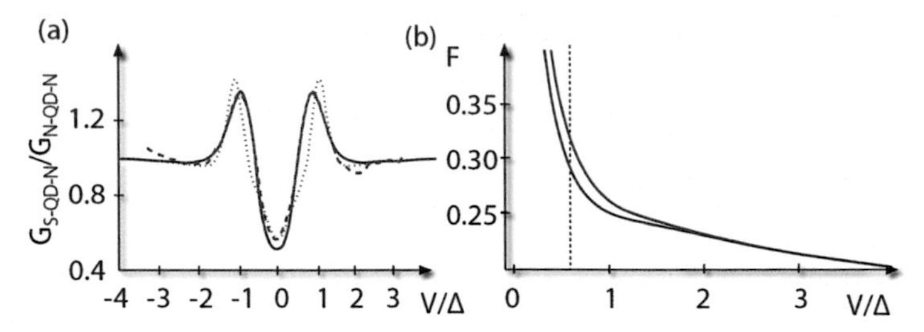


**Figure 6.7.:** **(a)**: theoretical prediction for the differential conductance $G_{\mathrm{S-QD-N}}$ divided by the normal state value $G_{\mathrm{N-QD-N}}$ shown as the solid curve. The parameters taken from (Gräber et al. [ 2004]) include$\Gamma_K = 0.7\Delta$, $\Gamma = 0.05$, $\Gamma_D = 0.053\Delta$, $T_g = 0.375$ and the width of the Lorentzian for the conductance in the KNQN case of $0.6\Delta$. To obtain the peaks at the right position for our fit we need to employ a slightly lower SC gap compared to the fit parameter $\Delta$ in the semiconductor model (Tinkham [ 1996]) (dotted curve) taken also from (Gräber et al. [ 2004])$\tilde{\Delta} = 0.7\Delta$. The result can be compared to the experimental data shown as the dashed curve (Gräber et al. [ 2004]).
**(b)**: theoretical prediction for the Fano factor in the normal-QD-SC junction with the same experimental parameters as used in the fit in (a). The solid curve has been obtained with ARs and the dotted curve is the prediction without AR ($T_{\mathrm{AK}} = 0$). One observes a clear difference between the two curves that amounts up to a difference of about 15% at $V/\Delta = 0.6$ (see dotted line).

full energy-dependence of the transmission coefficients derived in the Hamiltonian approach. The energy-dependent transmission coefficients lead to broader conductance peaks close to the SC gap while the explicit inclusion of AR leads to an enhanced conductance at low bias. For higher voltages the structure in the background DOS starts to play a role and deviations are expected.
Due to the low transparency of the contact AR is strongly suppressed and causes only a small contribution to the differential conductance at low voltages. However, according to our effective theory it should be observable by a noise measurement.
In Fig. 6.7 (b) the Fano factor including AR has been compared to the prediction without ARs ($T_{\mathrm{AK}} = 0$). One observes a clear difference in the Fano factors. This effect can be used to identify the presence of Andreev processes even for large interaction strengths.
The possibility of AR also raises the question whether one could be able to observe a positive cross-correlation in a three-terminal setup that has been investigated in Section 6.1 . As we can see from an RG analysis as e.g. in Shah and Rosch [ 2006] a single QD coupled to two normal leads and a SC is not a good device since one of the normal leads will form a Kondo singlet and the other will be weakly coupled. Instead, we have to go over to two QDs. From the results obtained here we can address the case of a SC coupled to two Kondo dots contacted by two normal leads at the same chemical potential (see Fig. 6.8 (a)).
In principle a positive cross correlation of the currents in the two normal leads could be inferred again from CPS as mentioned in Sections 6.1 and 6.2 . For simplicity here we only want to address

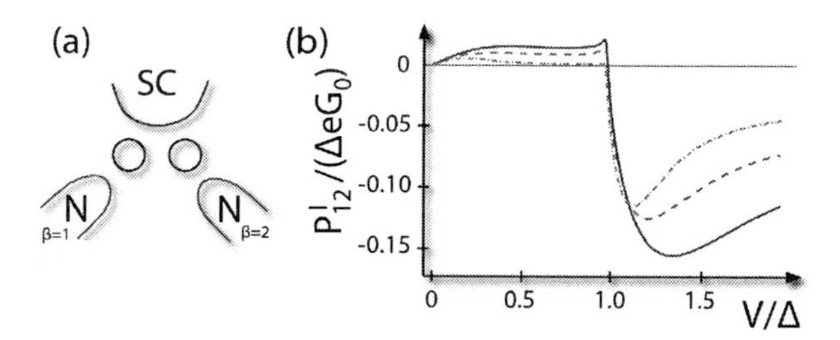


**Figure 6.8.:** **(a)**: sketch of the experimental setup for measuring the cross-correlation. A SC is contacted via two QDs in the Kondo regime to two normal leads (N) at the same chemical potential.
**(b)**: theoretical predicition for the cross-correlation $P^I_{12}$ at $T = 0$ in a three-terminal setup with one SC and two normal leads. All curves have been obtained for $\Gamma = 0.05$ and a broadening of the BCS DOS has been taken into account by a Dynes parameter of $\Gamma_D = 0.005\Delta$. The solid curve refers to $\Gamma_{K1} = 0.7\Delta = \Gamma_{K2}$, the dashed curve is for $\Gamma_{K1} = 0.3\Delta$, $\Gamma_{K2} = \Delta$ and the dotted curve has been obtained for $\Gamma_{K1} = 0.3\Delta = \Gamma_{K2}$.

the case of $T = 0$, where the CGF adopts the form (see also Eq. B.2)

$$\ln \chi_{\mathrm{SKK}}(\boldsymbol{\lambda}, \tau) = 2\tau \int_0^V \frac{d\omega}{\pi} \ln \left[1 + \sum_{\beta=1,2} T_{\mathrm{eK}\beta}(\omega)(e^{i\lambda_\beta} - 1)\right] \theta\left(\frac{|\omega| - \Delta}{\Delta}\right)$$
$$+\tau \int_{-V}^{V} \frac{d\omega}{\pi} \ln \left[1 + \sum_{\beta=1,2} T_{\mathrm{AK}\beta}(e^{2i\lambda_\beta} - 1) + T_{\mathrm{CAK}}(e^{i(\lambda_1+\lambda_2)} - 1)\right] \theta\left(\frac{\Delta - |\omega|}{\Delta}\right), \quad (6.49)$$

where the transmission coefficients can be inferred using Eqs. (B.4) and (B.5) to be

$$\begin{aligned}
T_{\mathrm{eK}\beta} &= \frac{4\Gamma_e}{(1+\Gamma_e)^2} \frac{\Gamma_{K\beta^2}}{(\omega - V)^2 + \Gamma_{K\beta}^2}, \\
T_{\mathrm{AK}\beta} &= \frac{4\Gamma_A^2}{\Gamma_A^4 + 2\Gamma_A^2(1-\Gamma_e^2) + (1+\Gamma_e^2)^2} \frac{\Gamma_{K\beta}^4}{[(\omega - V)^2 + \Gamma_{K1}^2][(\omega + V)^2 + \Gamma_{K2}^2]}, \\
T_{\mathrm{CAK}} &= \frac{8\Gamma_A^2}{\Gamma_A^4 + 2\Gamma_A^2(1-\Gamma_e^2) + (1+\Gamma_e^2)^2} \frac{\Gamma_{K1}^2\Gamma_{K2}^2}{[(\omega - V)^2 + \Gamma_{K1}^2][(\omega + V)^2 + \Gamma_{K2}^2]}.
\end{aligned}$$

The cross correlation can be calculated as a second derivative of the CGF

$$P^I_{12} = -\frac{1}{\tau} \frac{\partial^2 \ln \chi_{\mathrm{SKK}}(\boldsymbol{\lambda}, \tau)}{\partial\lambda_1 \partial\lambda_2}\bigg|_{\boldsymbol{\lambda}=0}. \quad (6.50)$$

Fig. 6.8 (b) shows the result for three different configurations of the widths of the respective Kondo resonances. In all cases we observe a positive cross-correlation that first increases towards the SC gap and then quickly disappears for voltages above the gap. This means that in principle it should be possible to observe a positive cross-correlation of the currents in the normal leads that is mediated by CPS. Above the SC gap single-electron transfer to the separate terminals sets in and causes a negative cross-correlation. However, we should mind that the prediction has been done for $T = 0$. At finite temperature thermally activated single-electron transfer becomes possible. This process causes a negative correlation of the two currents that will, in most cases, destroy the positive cross-correlation by CPS. Nevertheless, this shows that even in the Kondo regime a positive cross-correlation could in principle be observable.

## 6.5. Conclusions

We have investigated several aspects of Cooper pair splitters involving a SC in a hybrid configuration. In the non-interacting case we have discussed the principle of CPS and have additionally found the AET transport mechanism that also produces entangled electrons. Then, we inspected the effect of spin-active scattering in SC-FM hybrids and found that spin-active scattering can be used to create entangled $p$-wave pairs of electrons. The investigation of interaction effects led to the conclusion that the effect of interaction is not just to suppress charge transfer below the gap but to also give rise to new transport mechanisms: above the gap the well-known effect of coherent tunneling of electron pairs due to the onsite interaction is observed. Below the gap the same mechanism leads to coherent ARs giving rise to quartic terms in the CGF. Finally, we adressed the typical Kondo situation using an effective model and made contact to contemporary experiments. In order to observe CPS with QDs in the Kondo regime we had to use two QDs. Indeed, the larger degree of control in a double dot experiment makes such setups the most relevant ones in current experiments. We will treat this more involved setup in closer detail in the next Chapter 7.

# Chapter 7

## Double dot Cooper pair splitters

In Chapter 6 we introduced the principles of CPS. In Section 6.4 we have investigated a double dot Cooper pair splitter in a typical Kondo situation. In order to obtain an efficient EPR source we have seen already in Section 6.1 that the splitting of pairs into separate electrons has to be enforced. This can be achieved by the electrons 'repelling' each other by Coulomb interaction (Recher et al. [2001]). Controlled Cooper pair splitting can thereby be realized by coupling of the SC to two normal metal contacts via individually tunable QDs. Typically, these two are operated in an even charge state in contrast to the splitter in Section 6.4 .
The necessary double quantum dot devices have so far been realized either using InAs nanowires (Das et al. [ 2012], Hofstetter et al. [ 2009], Hofstetter et al. [ 2011]) or carbon nanotubes (Herrmann et al. [ 2010], Herrmann et al. [ 2012], Schindele et al. [ 2012]). In spite of recent theoretical progress (Burset et al. [ 2011], Rech et al. [ 2012]), the experimental data, especially of InAs based splitters, can only be modelled using the theory outlined in Soller et al. [ 2012a] that will be described below. We use the generic model for CPS from Burset et al. [ 2011] but also demonstrate that simpler models for specific experiments can often be found with the $T$-matrix approach (Recher et al. [ 2001]) or a master equation method (Eldridge et al. [ 2010]).

### 7.1. Generic model for the double quantum dot CPS device

A schematic of a CPS device is shown in Fig. 7.1 . A nanowire/carbon nanotube is used to form two QDs that contact a SC strip (S) in the center and two normal metal leads (N1, N2) at the edges. The two QDs can additionally be tuned via two uncoupled top gates.
Similar to Eq. (6.2) we use a tunneling Hamiltonian description of the system $H = H_0 + H_{\mathrm{TDD}}$ as in Recher et al. [ 2001], where

$$H_0 = H_2 + H_S + \sum_{\substack{i=1,2\\ \sigma=\uparrow,\downarrow}} \epsilon_{Di} d^\dagger_{i\sigma} d_{i\sigma} + \sum_{i=1,2} U_i n_{i\uparrow} n_{i\downarrow}. \tag{7.1}$$

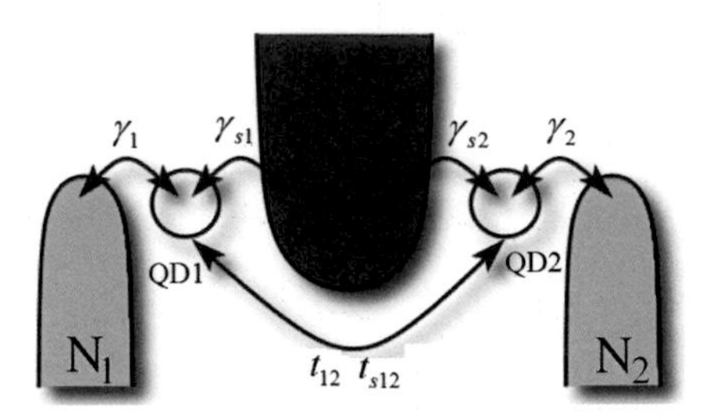


**Figure 7.1.:** Schematics of the generic model for a typical Y-junction: a SC is contacted via two QDs to two normal leads (N1, N2) possibly at a different chemical potential. The two QDs have tunable levels and may be coupled either via the superconductor ($t_{S12}$) or directly ($t_{12}$).

The dots are characterized by spin-degenerate resonant levels at $\epsilon_{Di}$ and charging energies $U_i$. The additional effective tunnel rates $t_{S12}$, $t_{12}$ couple the QDs either directly or via the SC. Following the method of Burset et al. [ 2011] we introduce them later in the bare double dot Hamiltonian and the self-energy describing the coupling with the SC electrode. $H_{\text{TDD}}$ corresponds then only to the tunnel coupling between the dots and the leads as in Eq. (6.4)

$$\begin{aligned} H_{\text{TDD}} &= H_{\text{NT1}} + H_{\text{NT2}} + H_{\text{ST1}} + H_{\text{ST2}} \qquad (7.2) \\ &= \sum_{\sigma} \gamma_1 \left( d_{1\sigma}^{+} \Psi_{1\sigma}(x=0) + \text{H.c.} \right) + \sum_{\sigma} \gamma_2 \left( d_{2\sigma}^{+} \Psi_{2\sigma}(x=0) + \text{H.c.} \right) \\ &+ \sum_{i=1,2,\,\sigma} \gamma_{DSi} \left( d_{i\sigma}^{+} \Psi_{i\sigma}(x=0) + \text{H.c.} \right). \qquad (7.3) \end{aligned}$$

The transport properties of this model can be obtained by calculating the respective GFs. In a combined dot-(electron/hole) space described by spinor fields $\Psi = \left( d_{1\uparrow}, d_{1\downarrow}^{\dagger}, d_{2\uparrow}, d_{2\downarrow}^{\dagger} \right)^T$ and in the linear response regime, we can characterize the model by a retarded GF

$$\hat{G}^r(E) = \left[ \omega - \hat{h}_0 + \hat{\Gamma}_N - \hat{\Sigma}^r(\omega) \right]^{-1}, \qquad (7.4)$$

where the symbol $\cdot\hat{\cdot}\cdot$ denotes a $4 \times 4$ matrix. Hereafter, we assume that $\omega$ stands for $\omega + i\eta$ and the limit $\eta \to 0^+$ is taken to obtain the retarded component of the GF. The Hamiltonian matrix describing the double dot isolated from the leads but taking into account the effective inter-dot tunnel rate $t_{12}$ is given by

$$\hat{h}_0 = \begin{pmatrix} \epsilon_{D1} & t_{12} \\ t_{12} & \epsilon_{D2} \end{pmatrix} \otimes \sigma_z, \qquad (7.5)$$

where $\sigma_{x,y,z}$ are the Pauli matrices in electron/hole-space. We describe the coupling with the normal leads as

$$\hat{\Gamma}_N = \begin{pmatrix} i\Gamma_1 & 0 \\ 0 & i\Gamma_2 \end{pmatrix} \otimes \begin{pmatrix} 1 & 0 \\ 0 & 1 \end{pmatrix}. \qquad (7.6)$$

The self-energy $\hat{\Sigma}^r(E) = \hat{\Gamma}_S(\omega) + \hat{\Sigma}_U(\omega)$ includes the Coulomb interactions ($\hat{\Sigma}_U$) and the coupling to the SC electrode ($\hat{\Gamma}_S$). Along with the effective inter-dot tunnel rate $t_{S12}$ via the SC the latter is given by

$$\hat{\Gamma}_S = \begin{pmatrix} \Gamma_{\text{DS1}} & t_{S12} \\ t_{S12} & \Gamma_{\text{DS2}} \end{pmatrix} \otimes \left[ g(\omega)\sigma_0 - f(\omega)\sigma_x \right], \qquad (7.7)$$

with $\Gamma_{DS1,DS2}$ the tunneling rates between the SC and each dot, $t_{S12}$ the inter-dot tunneling rate through the superconductor and $g(\omega) = -f(\omega)\omega/\Delta = -\omega/\sqrt{\Delta^2 - \omega^2}$ the dimensionless BCS retarded GFs of the uncoupled SC lead.
Once an expression for $\hat{G}^r(E)$ has been obtained, the linear transport coefficients can be computed. The contribution to the zero temperature linear conductance due to local AR processes at each dot is given by $R_{A1(2)}(\omega) = 4|\hat{G}^r_{e1,h1(e2,h2)}(\omega)|^2\Gamma^2_{1(2)}$. The probability of CPS is $T_{\rm CPS}(E) = 4|\hat{G}^r_{e1,h2}(\omega)|^2\Gamma_1\Gamma_2$ and that of EC is $T_{\rm EC}(E) = 4|\hat{G}^r_{e1,e2}(E)|^2\Gamma_1\Gamma_2$. Thus, the linear conductance at zero temperature at one of the normal electrodes, e.g. electrode 1, is given by $G_1(\omega) = G_0[R_{A1}(\omega) + T_{\rm CPS}(\omega) - T_{\rm EC}(\omega)]$, with $G_0 = 2e^2/h$.

### 7.1.1. Non-interacting case

Before taking into account Coulomb interactions, we want to discuss the effect of the effective tunnel rates between the QDs. We can tunnel between the two dots either directly or via the SC.
Of particular interest for determining the splitting performance of the device are the quantities $\Sigma_{e1,e2}$ and $\Sigma_{e1,h2}$, which correspond to the inter-dot CPS and EC processes. In the regime where $\Gamma_{1,2} \gg t_{12}, t_{S12}$, and in the absence of electronic interactions, we have that $\Sigma_{e1,h2} \propto t_{S12}$ and $\Sigma_{e1,h2} \propto t_{12}$. The inter-dot couplings in the case of a nanowire/nanotube in between the QDs as in Fig. 7.1 are effective parameters that have to be chosen depending on the specifics of the experiment in question. The transmission coefficients for CPS and EC are given by

$$T_{\rm EC}(\omega) = 4\Gamma_1\Gamma_2|G_{e1,he2}|^2 \overset{t_{12}\ll\Gamma_{1,1}}{\approx} \frac{4\Gamma_1\Gamma_2 t_{12}^2}{[(\omega-\epsilon_1)^2+\Gamma_1^2][(\omega-\epsilon_2)^2+\Gamma_2^2]}, \tag{7.8}$$

$$T_{\rm CPS}(\omega) = 4\Gamma_1\Gamma_2|G_{e1,h2}|^2 \overset{t_{S12}\ll\Gamma_{1,2}}{\approx} \frac{4\Gamma_1\Gamma_2 t_{S12}^2}{[(\omega-\epsilon_1)^2+\Gamma_1^2][(\omega-\epsilon_2)^2+\Gamma_2^2]}. \tag{7.9}$$

In this regime, the local ARs are proportional to $\Gamma^2_{DS1,DS2}$ and will be dominant.
Concerning the effective parameters $t_{S12}$ and $t_{12}$ we use two approaches: for the case of a carbon nanotube we follow the analysis of a tight-binding model in Herrera et al. [2010], where it was shown that these effective parameters are constant. The situation in the case of a semiconducting nanowire like InAs is different as we show in Appendix C.

### 7.1.2. Electronic interactions

The quantity $\hat{\Sigma}_U$ takes into account the effect of Coulomb interactions within the dots. To lowest order in $U$ this is given by a Hartree-Fock approximation (Cuevas et al. [2001]) as $\hat{\Sigma}_U = U_i\langle n_i\rangle\sigma_z + \Delta_i\sigma_x$, see Vecino et al. [2003], where $\Delta_i = U_i\langle d^\dagger_{i\uparrow}d^\dagger_{i\downarrow}\rangle$ is related to the proximity effect induced order parameter in each dot as we discussed in Section 6.3.2. As far as Kondo correlations can be neglected, i.e., when the conditions are such that $T > T_K$, where $T_K$ is the Kondo temperature, this term has the effect of reducing the amplitude of ARs on each dot by renormalizing the couplings to the superconductor. For a more detailed discussion we refer to Atienza [2012]. An extension of this scheme to second order in $U$ and beyond is straightforward (Vecino et al. [2003]).
Since typically CPS experiments focus on QDs on resonance so that no Kondo correlations are expected, we can neglect the effect of Coulomb interactions keeping in mind that the bare position and width of the resonance is additionally affected by the interactions leading to a renormalised dot position and resonance width (Chevallier et al. [2011]).

## 7.2. Finite bias experiments

In the finite bias measurement in Hofstetter et al. [2011], first, the local zero bias conductance $G_1$ between the SC and the first normal lead (N1) is measured for varying gate voltages $U_{\mathrm{g1}}$ around a resonance (see Fig. 2.3 (a)) as well as the voltage $U_{\mathrm{N2}}$ applied on the second lead relative to N1 and the SC. The main contribution to conductance comes from local transport processes between the SC and N1. Hofstetter et al. [2011] call the conductance fraction of $G_1$ which depends on $U_{\mathrm{N2}}$ non-local conductance. All non-local processes become ineffective at $|U_{\mathrm{N2}}| \gg \Delta/e$, thus the non-local conductance $\Delta G_1$ is obtained from the experimental data by subtraction: $\Delta G_1(U_{\mathrm{g1}}, U_{\mathrm{N2}}) = G_1(U_{\mathrm{g1}}, U_{\mathrm{N2}}) - G_1(U_{\mathrm{g1}}, U_{\mathrm{N2}} = -1\ \mathrm{mV})$. $|U_{\mathrm{N2}}| = 1$ mV satisfies the requirement $|U_{\mathrm{N2}}| \gg \Delta/e$. The first observation from the experimental data is that varying the bias on the second dot leads only to small variations in the local conductance $G_2$. We describe this feature by the dot energy level $\epsilon_2$ being pinned to the chemical potential $U_{\mathrm{N2}}$ of the second lead. Such pinning can either be caused by a Kondo resonance or simply by the gate not being fully operational so that it works as aligning the dot level with the chemical potential of the lead.
In Appendix C we calculate approximate expressions for the non-local CPS and EC conductance using the $T$-matrix approach in Eqs. (C.6) and (C.13). Including the pinning of the dot resonance of the second dot to $eU_{\mathrm{N2}}$ and the dependence on the gate voltage $U_{\mathrm{g1}}$ we arrive at the two expressions

$$\tilde{G}_{\mathrm{CPS}} = \frac{4e^2}{h}\frac{M_{\mathrm{CPS}}\Gamma^4_{\mathrm{CPS}}}{\{[\alpha(eU_{\mathrm{g1}}-\Delta_g)+eU_{\mathrm{N2}}]^2+\Gamma^2_{\mathrm{CPS}}\}^2}\rho(eU_{\mathrm{N2}}), \tag{7.10}$$

$$\tilde{G}_{\mathrm{EC}} = \frac{4e^2}{h}\frac{M_{\mathrm{EC}}\Gamma^2_{\mathrm{EC1}}\Gamma^2_{\mathrm{EC2}}}{[\alpha^2(eU_{\mathrm{g1}}-\Delta_g)^2+\Gamma^2_{\mathrm{EC1}}][(eU_{\mathrm{N2}})^2+\Gamma^2_{\mathrm{EC2}}]}\rho(eU_{\mathrm{N2}}), \tag{7.11}$$

where $\Delta_g$ refers to the resonance position of the first dot, $\Gamma_{\mathrm{CPS}}$, $\Gamma_{\mathrm{EC1}}$, $\Gamma_{\mathrm{EC2}}$ are the resonance widths for CPS and EC via the first/second dot and $\rho(x) = \theta(\Delta - x)\theta(\Delta + x)$. $M_{\mathrm{CPS}}$ and $M_{\mathrm{EC}}$ are matrix elements describing strength of the CPS and EC charge transfer processes. Additionally, we include the lever arm of the top gate $\alpha$ determined in the experiment to be $\alpha \approx 0.3$.
At finite temperature we have to integrate using the appropriate Fermi distributions

$$\tilde{G}_{\mathrm{CPS}}(U_{\mathrm{N2}}, T) = \frac{\partial}{\partial U_{\mathrm{N1}}}\int \frac{d\omega}{2\pi}\tilde{T}_{\mathrm{CPS}}(\omega)[n_2(\omega) - n_F(\omega - U_{\mathrm{N1}})]\bigg|_{U_{\mathrm{N1}}=0}, \tag{7.12}$$

$$\tilde{G}_{\mathrm{EC}}(U_{\mathrm{N2}}, T) = \frac{\partial}{\partial U_{\mathrm{N1}}}\int \frac{d\omega}{2\pi}\tilde{T}_{\mathrm{EC}}(\omega)[n_2(\omega) - n_F(\omega - U_{\mathrm{N1}})]\bigg|_{U_{\mathrm{N1}}=0}, \tag{7.13}$$

where $\tilde{T}_{\mathrm{EC}}$ and $\tilde{T}_{\mathrm{CPS}}$ are given by Eqs. (7.10) and (7.11), using the substitutions $U_{\mathrm{N2}} \to \omega$ and $n_2(\omega) = n_F(\omega - U_{\mathrm{N2}})$.
The non-local conductance is now given by

$$\Delta G_1 = \tilde{G}_{\mathrm{CPS}}(U_{N2}, T) - \tilde{G}_{\mathrm{EC}}(U_{N2}, T). \tag{7.14}$$

This approximate treatment can be compared to the experimental data obtained in Hofstetter et al. [2011], in order to verify that this treatment is capable of obtaining the correct energy dependence of $t_{S12}$ and $t_{12}$.
For the local conductance $G_1$ we take Eq. (6.9). In the interacting case the same formula can be recovered (Cuevas et al. [2001]), however, with renormalised parameters $\epsilon_D$ and $\Gamma_{\mathrm{DS1}}$ involving the respective 11- and 12-component of the Nambu self-energy.
We fit the parameters to the experimental data and obtain $M_{\mathrm{CPS}} = 0.195$, $M_{\mathrm{EC}} = 0.095$, $\Gamma_{\mathrm{CPS}} = 0.25$ meV, $\Gamma_{\mathrm{EC1}} = 0.21$ meV and $\Gamma_{\mathrm{EC2}} = 0.27$ meV. We observe good agreement between the

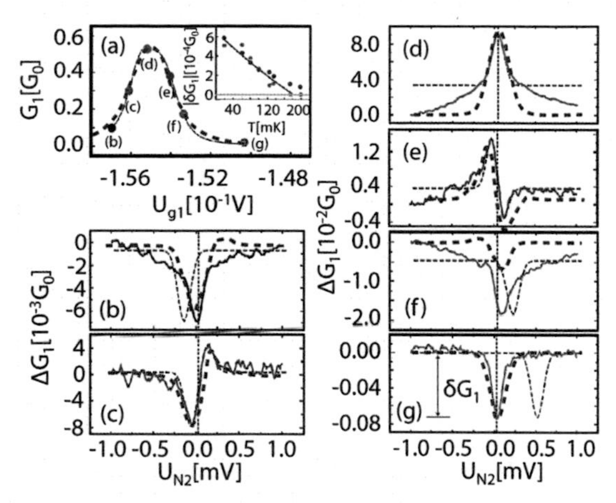


**Figure 7.2.:** Bias dependence of $\Delta G_1$ for a series of top gate voltages $U_{g1}$ at $T = 20$ mK. The latter are indicated in (a) where $G_1$ is shown and compared to the theoretical prediction for $\Gamma_1 = 0.12$ meV and $\Gamma_{DS1} = 0.63$ meV using Eq. (6.9). The color code in (a) indicates the gate voltage in (b)-(g). The thin dashed curves are derived from the model discussed in Hofstetter et al. [ 2011] and the dashed blue curves are derived from the model discussed in the text. Vertical lines indicate $U_{N2} = 0$, including a small offset from the $I-V$-converter on N2. The inset in (a) shows the temperature dependence of a minimum close to the one in (g) with a black line as a guide to the eye. The blue dots are the result of the theoretical model. We use parameters $M_{CPS} = 0.195$, $M_{EC} = 0.095$, $\Gamma_{CPS} = 0.25$ meV, $\Gamma_{EC1} = 0.21$ meV and $\Gamma_{EC2} = 0.27$ meV.

theoretical prediction and the experimental data with the exception of curve (f). Comparing our result with the model from Hofstetter et al. [ 2011] we see that we correctly capture the position of all peaks and dips. We also obtain the correct peak heights without using additional offset conductances. As in the experimental work we find that $M_{CPS} \gg M_{EC}$ so that on resonance CPS is dominant. However, CPS depends on the sum of both level positions, whereas EC only depends on the product of the two Lorentzians associated to the two QDs. Therefore, EC becomes dominant off resonance as the CPS peak is reduced outside the SC gap.
Concerning the fitting parameters we found $\Gamma_{EC2}$ should be equal to $\Gamma_2$ being the tunnel rate between the second QD and the second drain. $\Gamma_{EC2} = 0.27\,\text{meV}$ is a reasonable value. From our model above we also concluded that $\Gamma_{CPS} = \Gamma_1 + \Gamma_2$, which is at least partially fulfilled: $0.25\,\text{meV} \approx 0.12\,\text{meV} + 0.27\,\text{meV} = 0.39\,\text{meV}$. Also, $\Gamma_{EC1} = 0.21\,\text{meV} \approx \Gamma_1 = 0.12\,\text{meV}$ follows approximately the model derived above. A broadening of the resonances, as observed here, can be attributed to higher order tunneling processes between the SC and the QDs which were neglected in the non-local processes.
Compared to the model used in Hofstetter et al. [ 2011] two effects are accounted for in our model: first, we take into account the effect of interactions and the geometric suppression which both lead to different tunneling rates to the SC for the local and non-local processes. Second, we account for the energy-dependent SC gap leading to a suppression of the non-local processes for voltages larger than $\Delta$.
Next, we consider the temperature dependence of the conductance peaks. Apart from the Fermi distribution in Eqs. (7.12) and (7.13) we include the temperature dependence of the gap, which is

assumed to be given by the Thouless equation (Thouless [ 1960])

$$\frac{\Delta(T)}{\Delta} = \tanh\left(\frac{\Delta(T)\cdot T_c}{\Delta\cdot T}\right), \tag{7.15}$$

where $\Delta$ refers to the bare (zero-temperature) gap and $\Delta(T)$ is the gap at finite temperature. $T_c$ is the critical temperature known from BCS theory $k_B T_c = \Delta/1.76$.
We obtain from the measured SC gap in the experiment $T_c \approx 850$ mK. In the inset of Fig. 7.2 (a) the temperature dependence of the EC dip close to the one shown in Fig. 7.2 (g) is shown. We study the temperature dependence of $\Delta G_1$ from Eq. (7.14) using Eqs. (7.12) and (7.13).
We observe a rapid decrease and broadening of the EC dip for increasing temperatures. The peak positions for the dips at different temperatures are also compared to the experimental data in the inset of Fig. 7.2 (a) and good agreement is obtained. Especially, we reproduce a linear decrease of the signal due to the broadening of the Fermi distribution. We note that the EC dip does not disappear completely in our theoretical model but it becomes very broad and shallow so that it almost disappears around the temperature $T \approx 175$ mK as in the experiment. This tempeature is much smaller than $T_c$ and indeed we find that the SC gap only changes by 2% for the temperatures considered here. The same behavior has been observed in a different experiment (Hofstetter et al. [2009]), where the authors concluded that the non-local signal is not controlled by the bulk $\Delta$ alone. The model developped here allows for an explanation of the temperature dependence encountered in the experiments: the non-local signal is mainly controlled by the temperature dependence of the distribution functions and not by the temperature dependence of the SC gap.
We conclude that the approximate treatment presented in Appendix C works well. The results can be reproduced by the model in Section 7.1 choosing

$$t_{12} = \Gamma_{S1}\Gamma_{S2}\left[\frac{\cos(k_F\delta r)}{k_F\delta r}\right]^2 \exp\left(-\frac{2\delta r}{\pi\xi}\right), \tag{7.16}$$

$$t_{S12} = \frac{\Gamma_{S1}\Gamma_{S2}[(\omega-\epsilon_{D1})^2+\Gamma_1^2][(\omega-\epsilon_{D2})^2+\Gamma_2^2]}{[(\omega-\epsilon_{D1}-\epsilon_{D2})^2+(\Gamma_1+\Gamma_2)^2]^2}\left[\frac{\sin(k_F\delta r)}{k_F\delta r}\right]^2 \exp\left(-\frac{2\delta r}{\pi\xi}\right), \tag{7.17}$$

where $k_F$ is the Fermi velocity in the SC and $\xi$ is the SC coherence length.
Finally, we also investigate the case of a constant $t_{S12}$ using the model in Section 7.1 , which would correspond to an experiment using carbon nanotubes. Again we try to reproduce the behavior of the non-local conductance as indicated in Fig. 7.2 , see Fig. 7.3 . Zero-bias conductance at small temperature is directly related to the abovementioned transmission coefficients via

$$G_{\mathrm{EC}}(U_{\mathrm{g1}}, U_{\mathrm{N2}}) = T_{\mathrm{EC}}(0)_{\epsilon_{D1}=U_{\mathrm{g1}},\epsilon_{D2}=U_{\mathrm{N2}}}, \tag{7.18}$$

$$G_{\mathrm{CPS}}(U_{\mathrm{g1}}, U_{\mathrm{N2}}) = T_{\mathrm{CPS}}(0)_{\epsilon_{D1}=U_{\mathrm{g1}},\epsilon_{D2}=U_{\mathrm{N2}}}. \tag{7.19}$$

The voltage $U_{\mathrm{g1}}$ additionally has to be rescaled with the lever arm of the top gate. In Fig. 7.3 we observe the same qualitative behavior as in the experiment, however, we have not reached quantitative agreement. This can be reached only incorporating the aforementioned energy-dependence of $t_{S12}$ from Eq. (7.17). Nonetheless, we still reproduce the qualitative behavior without this specific energy-dependence and we may therefore expect that a similar behavior could also be observed in an experiment involving carbon nanotubes.

## 7.3. Efficiency of CPS

In this Section we turn again to carbon nanotube based splitters as used in a recent experiment (Schindele et al. [ 2012]), where the aim was to optimize the splitting efficiency. In this experiment

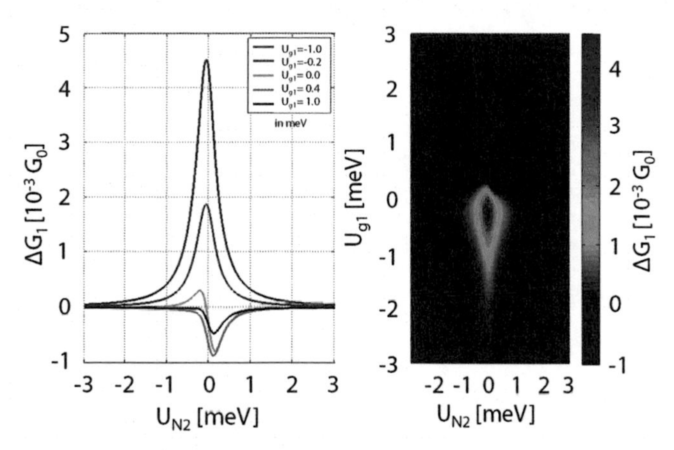


**Figure 7.3.:** Left: plot of the non-local conductance $G_{\text{CPS}} - G_{\text{EC}}$ at zero energy and temperature as a function of $U_{\text{N2}}$ for different values of $U_{\text{g1}}$ calculated using the model in Section 7.1 . The tunneling rates are$\Gamma_1 = 0.6$ meV, $\Gamma_2 = 0.3$ meV, $\Gamma_{\text{DS1}} = 0.03$ meV, $\Gamma_{\text{DS2}} = 0.005$ meV, $t_{S12} = 0.03$ meV, $t_{12} = 0.029$ meV. The SC gap is $\Delta = 0.13$ meV. Right: color map of the non-local conductance as a function of the gate $U_{\text{g1}}$ and $U_{\text{N2}}$ for the same parameters.

zero bias conductances between N1 and the SC ($G_1$) and N2 and the SC ($G_2$) have been recorded as a function of the side gate voltages on the first and second quantum dot $U_{\text{sg1}}$ and $U_{\text{sg2}}$. In the experiment two features are observed: a typically broad local conductance peak due to local pair tunneling (LPT) and a narrower additional peak due to CPS. LPT refers to the break-up of a Cooper pair due to the finite SC gap, which leads to a subsequent transfer of two electrons via one of the QDs. The additional CPS peaks vanish when superconductivity is suppressed using an additional magnetic field.

In order to access the non-local conductances $\Delta G_1$ and $\Delta G_2$ the total conductance (e.g. $G_1$) as a function of one gate voltage (e.g. $U_{\text{sg2}}$) is measured and the signal is compared to a Lorentzian which can be expected if only the local conductance would contribute. The excess conductance is then defined as $\Delta G_1$. From the experiment it is suggested that such procedure corresponds to comparing $G_1$ in presence of the additional coupling $t_{S12}$ and without it, see Fig. 7.1 .

One of the most remarkable features of the experimental data is that $\Delta G_1 \neq \Delta G_2$, whereas one would naively expect $\Delta G_1 = \Delta G_2$. This discrepancy is reproduced when calculating the total conductance using the model in Section 7.1 using a sizeable$t_{S12}$.

The observed discrepancy can easily be explained: changing $t_{S12}$ from a finite value to zero does not only eliminate CPS but also changes the DOS on the two QDs. Changing the DOS also affects the local conductances so that $\Delta G_1$ and $\Delta G_2$ do not only represent CPS but also the change of the local conductances. However, measuring $G_1$ and $\Delta G_1$ allows for a precise determination of $t_{S12}$ so that one may calculate the true CPS conductance from our model.

In Appendix D we show how to go over from the generic model to a simple master equation and how to reproduce the abovementioned discrepancy. In addition, we show that LPT gives rise to a finite amplitude for triplet splitting.

The different complications of the measurement scheme used in (Schindele et al. [ 2012]) in order to extract the CPS conductance stem from the fact that setting $t_{S12} = 0$ also changes the DOS of the QDs which causes manifold difficulties when trying to identify the CPS conductance. We believe that these complications are also the reason for the peculiarities that were encountered in a similar measurement for an InAs based splitter (Das et al. [ 2012]), where splitting efficiencies larger than 100% have been reported.

## 7.4. Cross-correlation

Based on the good agreement with the data for the InAs splitter in Section 7.2 we use the presented model to show how to access also the cross-correlation of currents as in Sections 6.1 and 6.2 .
We introduce two transmission coefficients that describe the local processes in the SC beamsplitter. The first one is due to the local charge transfer between the SC and N1, where no bias is applied so that $T_1 = G_1(\epsilon_D \to \epsilon_D - \omega)$. The second is due to local charge transfer between N2 and the SC. In the experiment for the second QD a very broad resonance was found (Hofstetter et al. [ 2011]). Since we have assumed this to be due to the dot resonance being pinned to the Fermi level of the second lead, the transmission coefficient will be constant. Furthermore, we assume charge transfer between N2 and the SC to be only due to tunneling of single electrons which is a conservative estimate as will become clear below. The amplitude of the resonance is assumed to be the same as for the first QD so that we introduce

$$T_1 = \frac{G_1(\epsilon_D \to \epsilon_D - \omega)}{2}, \; T_2 = \frac{2e^2}{h} M_2, \tag{7.20}$$

where $M_2$ corresponds to the height of the second resonance. The exchange in $T_1$ has to be done in order to obtain a properly normalised transmission coefficient.
The calculation of cross-correlations can be done straightforwardly (Rech et al. [ 2012]) but involves lengthy expressions. We want to take a different approach here by calculating directly the CGF, which provides direct access to all higher order cumulants via the CGF. We have calculated the CGF for a SC beamsplitter already in Section 6.1 , however, for a slightly different geometry. Nonetheless, no other charge transfer processes but the ones already identified in Section 6.1 are taking place so that we can use the approach presented in Komnik and Saleur [ 2006]: we take the structure of the CGF in Eq. (B.2) and use the transmission coefficients from the model described in Section 7.2 . This way the cumulant generating function $\chi_{\mathrm{DD}}(\boldsymbol{\lambda})$ reads

$$\begin{aligned} \ln \chi_{\mathrm{DD}}(\boldsymbol{\lambda}) \;=\; & \tau \int_{-\infty}^{\infty} \frac{d\omega}{\pi} \ln \Big\{ 1 + T_1[(e^{i2(\lambda_1-\lambda_S)} - 1) n_F(1-n_F) + (e^{-i2(\lambda_1-\lambda_S)} - 1) n_F(1-n_F)] \\ & + T_2[(e^{i(\lambda_2-\lambda_S)} - 1) n_2(1-n_F) + (e^{-i(\lambda_2-\lambda_S)} - 1) n_F(1-n_2)] \\ & + \tilde{T}_{\mathrm{CPS}}(\omega)[(e^{i(\lambda_1+\lambda_2-2\lambda_S)} - 1) n_2(1-n_F) + (e^{-i(\lambda_1+\lambda_2-2\lambda_S)} - 1) n_F(1-n_2)] \\ & + \tilde{T}_{\mathrm{EC}}(\omega)[(e^{i(\lambda_2-\lambda_1)} - 1) n_2(1-n_F) + (e^{-i(\lambda_2-\lambda_1)} - 1) n_F(1-n_2)] \Big\}. \end{aligned} \tag{7.21}$$

$\boldsymbol{\lambda} = (\lambda_1, \lambda_2, \lambda_S)$ refers to the counting fields for the first, second and SC lead and we have used the transmission coefficients from Eq. (7.20). From the CGF we calculate the cross-correlation of the currents through the first and second lead via

$$P_{12}^I = -\frac{1}{\tau} \frac{\partial^2 \ln \chi_{\mathrm{DD}}(\boldsymbol{\lambda})}{\partial \lambda_1 \partial \lambda_2} \bigg|_{\boldsymbol{\lambda}=0}. \tag{7.22}$$

The result is shown in Fig. 7.4 for two temperatures $T = 20$ mK and $T = 200$ mK. We observe a positive cross-correlation in a small interval around the CPS resonance, which shows the importance

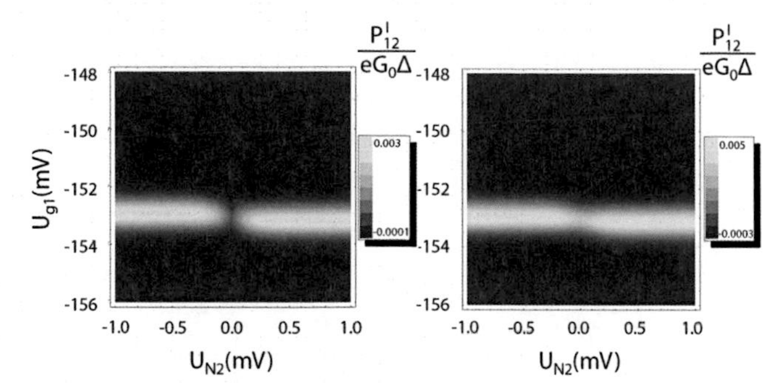


**Figure 7.4.:** Cross-correlation of the currents in the first and second drain calculated from Eq. (7.22) using parameters $M_{\mathrm{CPS}} = 0.195$, $M_{\mathrm{EC}} = 0.095$, $\Gamma_{\mathrm{CPS}} = 0.25\,\mathrm{meV}$, $\Gamma_{\mathrm{EC1}} = 0.21\,\mathrm{meV}$, $\Gamma_{\mathrm{EC2}} = 0.27\,\mathrm{meV}$, $M_2 = 0.55$, $\Gamma_1 = 0.12\,\mathrm{meV}$ and $\Gamma_{S1} = 0.63\,\mathrm{meV}$ as discussed before. The left density plot has been done for $T = 20\,\mathrm{mK}$ whereas the right one has been obtained for $T = 200\,\mathrm{mK}$.

of positive cross-correlation as a signature of CPS. At a larger temperature the positive cross-correlation is enhanced due thermally excited CPS processes. For top gate voltages away from resonance we observe a negative cross-correlation due to EC.
The general picture is therefore identical to other treatments of SC beamsplitters. However, we note that the overall value of the cross-correlations is very small. This is due to the fact that the local processes included in the cumulant generating function in Eq. (7.21) are involved in the denominator of Eq. (7.22) which means that they do not determine the sign but the overall value of the cross-correlations. This also explains why the precise form of the transmission coefficient and the type of charge transfer for the local processes described by the transmission coefficients in Eq. (7.20) are not of prior importance here. The large resonance of both transmission coefficients lead to a strong reduction of the cross-correlations mediated by CPS and EC. In this respect measuring the non-local conductance is advantageous since the local processes can be neglected completely.

## 7.5. Conclusions

To conclude, we have discussed how to describe Cooper pair splitters based on InAs and carbon nanotubes by a generic model. This model includes effective parameters which depend on the choice of materials in the experiment and have to be chosen accordingly. We have compared our predictions to recent experimental data and good agreement has been reached. We have shown how the two experimental methods used so far for extracting the non-linear conductance differ. In this way we have explained several recent observations. We have also demonstrated how to access the cross-correlation of currents and obtained a positive cross-correlation from CPS. However, the signal in the case of the InAs splitter discussed here is very small as it is affected also by the local processes. Therefore, we conclude that measuring the non-local conductance can be advantageous compared to the cross-correlation since it allows to neglect the local processes completely.
The characteristics of finite bias CPS are similar but not identical for carbon nanotube based and InAs nanowire based splitters. We have attributed the different behavior to the different energy dependence of the CPS transmission coefficient.

# Chapter 8

## Ferromagnet-quantum dot-superconductor junctions

In Sections 6.1 and 6.2 we already discussed several aspects of SC-FM hybrids involving a QD. The most promising realisation of such hybrids are based on InAs nanowires (see Section 2.3), since they allow for good contact to both constituents. Indeed, so far the experiment of Hofstetter et al. [2010] is the only experimental realisation. Additionally, the observed large $g$-factors of InAs nanowires have their origin in the large spin-orbit interaction which is fundamental for the rich field of spin physics.
We should therefore expect the system to show specific effects due to the FM correlations and the large $g$-factor both in the Kondo situation (that we analysed for a normal contact in Section 6.4) and for a QD on resonance (as in Chapter 7). A theoretical analysis of both regimes of such devices has so far only been done in Soller et al. [2012d] and we will follow its argumentation below.

## 8.1. FM-QD-SC device in the Kondo limit

First, we would like to show how to adopt the effective model for the Kondo limit described in Section 6.4 to the case of a FM terminal. In the experiment (Hofstetter et al. [2010]) the situation is such that one is clearly in the Kondo regime for an odd number of electrons on the QD (De Franceschi et al. [2010]). In this case one observes the many-body spin 1/2 Kondo resonance for temperatures below the Kondo temperature $T_K$ as described in Section 2.4. As in Section 6.4 SC and FM correlations are best visible in the case $T_K \lesssim \Delta$, where one observes a strong suppression of the effective hybridisation between the QD and the SC drain (Gräber et al. [2004]). In this case we can use the effective model described in Section 6.4, where the resulting transmission coefficients for the Kondo resonance are just the product of the transmission coefficient for the tunneling case discussed in Section 3.2 with the DOS of a resonant level, as we can also see from our calculation for the resonant level in Section 6.1.
Furthermore, one has to take into account that the Kondo resonance splits into a doublet due to the Zeeman effect and is also strongly affected by the exchange field of the FM. Using Haldane's scaling method for a flat band structure with spin-dependent tunneling rates and including a finite Stoner splitting of the leads an analytical formula for the energy splitting of the spin-$\uparrow$ and spin-$\downarrow$

bands is found to be (Martinek et al. [2005])

$$\delta_{\text{split}} = g\mu_B B + \Delta_s + \frac{P\Gamma_K}{\pi} \ln\left(\frac{|\epsilon_D|}{|U+\epsilon_D|}\right), \tag{8.1}$$

where $\Gamma_K$ is the hybridisation of the FM drain with the Kondo dot and $\epsilon_D$ is the position of the energy level of the QD. $g\mu_B B$ is the Zeeman splitting and $\Delta_s$ is a Stoner splitting induced shift (Sindel et al. [2007]). Eq. (8.1) is supported by a refined analysis based on numerical renormalisation group calculations (Martinek et al. [2005]). The two spin bands refers to a spin splitting of the Kondo resonance which can be described by an effective DOS with a Lorentzian shape given by

$$\rho_{K\sigma}(\omega) = \frac{\Gamma_K^2}{(\omega - eV + \sigma\delta_{\text{split}})^2 + \Gamma_K^2}. \tag{8.2}$$

This form assumes that we have a spin-splitted Kondo resonance that leads to two perfectly transmitting channels associated with the separate spin species. In the case of a FM lead spin-symmetry is broken so that the Kondo screening clouds associated to the two resonances are different and thus so are the couplings to the SC. In a first approximation they are given by the tunnel couplings in the SC-FM QPC. The transmission coefficients for the case of a SC-QD-FM device can consequently be deduced from the transmission coefficients for the SC-FM QPC (see Appendix B.1) and the relevant effective DOS as in Eq. (6.48)

$$T_{\text{eK}\sigma}(\omega) = T_{e\sigma}(\omega)\rho_{K\sigma}(\omega), \quad T_{\text{AKF}}(\omega) = T_{\text{AF}}(\omega)\rho_{K\sigma}(\omega)\rho_{K-\sigma}(-\omega). \tag{8.3}$$

We include the background conductance using the standard Levitov-Lesovik formula as in Eq. (6.47). The CGF for the FM-QD-SC junction is given by

$$\ln\chi_{\text{F-QD-S}} = \ln\chi_{\text{res}} + \ln\chi_g, \tag{8.4}$$

where $\chi_{\text{res}}$ can be derived from Eq. (B.1) by replacing $T_{\text{eF}\sigma}$ by $T_{\text{eK}\sigma}$ and $T_{\text{AF}}$ by $T_{\text{AKF}}$. In principle, one would also have to take into account branch-crossing and AR above the gap. However, we checked that in the limit of $\Gamma \ll 1$, which represents the typical experimental case of small hybridisation of the SC with the Kondo resonance, the corresponding transmission coefficients $T_{\text{AF2}}$ and $T_{\text{BCF}}$ in Eq. (B.1) may safely be neglected since their contribution is marginal.

We compare our results to the experimental data obtained in Hofstetter et al. [2010]. In the experimental setup a FM drain is formed by a Ni/Co/Pd trilayer and a Ti/Al bilayer is used as a SC drain. A QD forms in an InAs nanowire segment contacted by the FM and the SC. In agreement with previous experiments (Csonka et al. [2008]) the QD is perfectly controllable by a backgate voltage. The choice of InAs is essential as its $g$-factor in a wire geometry is comparable to the (rather big) bulk value (Csonka et al. [2008]).

Within this experiment the observed splitting of the Kondo resonance according to Eq. (8.1) has been verified. To test our model for the transport characteristics we choose a charge state that exhibits a clear signature of FM correlations, i.e. the Kondo resonance has a finite and roughly constant splitting at $B = 0$ T. We calculate the differential conductance and show the comparison of theory and experiment in Fig. 8.1.

We observe acceptable agreement in the voltage range considered here. Especially, we see that our model correctly describes the asymmetry of the two Kondo conductance peaks that originates from the different DOS for the two spin species. Concerning the fitting parameters of our model we find that we do not observe any quasiparticle-lifetime broadening of the SC DOS and the value for the polarisation is typical for Co based junctions as we have also seen from our analysis in Section 3.3. The width of the Kondo resonance has a typical size, comparable to the one found in Section 6.4.

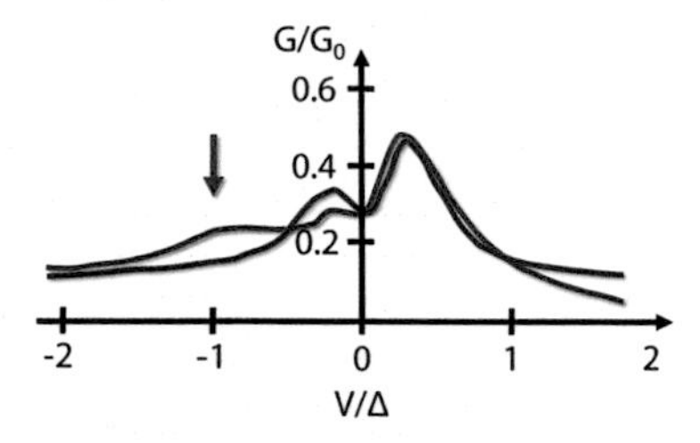


**Figure 8.1.:** Theoretical differential conductance (dashed curve) through the FM-QD-SC junction for $T = 0.13\Delta$, $\Gamma = 0.036$, $\Gamma_K = 0.29\Delta$, $P = 0.46$, $\delta_{\text{split}} = 0.16\Delta$ and $T_g = 0.035$. For the SC gap in Eq. (8.4) we use $\tilde{\Delta} = 0.23\Delta$, where $\Delta$ is the measured bare SC gap. One observes the characteristic double peak structure (Gräber et al. [ 2004]), however, now with the asymmetry related to the Kondo peak splitting. We compare our prediction to the experimental data (solid curve) taken from (Hofstetter et al. [ 2010]) at the background voltage $V_{BG} = 1.28$ V and $\Delta = 0.14$ meV. The red arrow indicates the bare SC gap observable in the experimental data.

From the fit we find $\Gamma_K \lesssim \Delta$, and thus $T_K \lesssim \Delta$ so that we access the interesting Kondo regime where the Kondo effect and superconductivity are concurring phenomena. The value $\delta_{\text{split}}$ allows for the calculation of the $g$-factor for the considered charge state since the critical magnetic field $B_c$ is related to the exchange field splitting via $g\mu_B B_c = 2\delta_{\text{split}}$. $B_c$ has been measured in the experiment to be 64 mT. The corresponding $g$-factor is 12, which is in perfect accordance with previous experimental studies of InAs nanowires (Csonka et al. [ 2008]). The small value$\tilde{\Delta} = 0.23\Delta$ signifies that the Kondo resonance couples to quasiparticle states within the SC gap, which can be ascribed to the granularity of the metallic contacts (Dynes et al. [ 1984]) and/or a nonzero DOS in the nanowire sections adjacent to the QD (Doh et al. [ 2008]). The most important source of deviations from our model is that we have neglected a possible energy dependence of the background DOS and the SC correlations on the QD. Indeed, due to the latter assumption in our model we do not see the bare SC gap at $V = -\Delta$ (see red arrow in Fig. 8.1 ).
The result reveals two basic facets of FM-QD-SC junctions. The first observation is that the background DOS (described by the transmission coefficient $T_g$) is very small. The second intriguing feature is that we did not have to include a spin-active tunneling term as in Eq. (3.24). Such a term would couple the tunnel transmission for one spin species to the Kondo singlet for the opposite spin, which would reduce the asymmetry in the peaks. Furthermore, SAR processes would have to be taken into account that couple only to one Kondo singlet and would therefore lead to a pronounced subgap feature. This effect will be discussed in Section 8.3 , where we show its relevance for a QD in the even state. Both phenomena are not observed and the value for the polarisation ($P = 0.46$), that reflects the asymmetry of the peaks, is in perfect accordance with previous experimental studies of point contacts. The irrelevance of spin-active scattering, even in the presence of a strongly polarized FM, is related to the strong asymmetry of the couplings between the dot and the FM or the SC respectively. The Kondo effect is mainly due to hybridized FM bulk states so that specifics of the interface or the SC are hardly seen. This also explains why the theory in Martinek et al. [2005] applies also for the case of a FM-QD-SC junction even though it has been derived for a QD coupled to two FM leads.
Let us finally also discuss the effects of AR. Due to the low tunnel coupling of the SC to the QD it is strongly suppressed. In Section 6.4 it was found that the presence of AR for a normal-QD-SC junction can be decided by a noise measurement. In the case of a FM-QD-SC junction the Fano factor does not change considerably since ARs are not only suppressed by the small tunnel coupling

but also by the Kondo peak splitting. Therefore, higher order cumulants, which are more sensitive to the doubled electron charge processes, are necessary to decide the presence of ARs in these devices.
It is remarkable that the asymmetry and splitting deduced from the model can be explained with a reasonable choice of the $g$-factor of InAs and the polarisation of Co. Additionally, the asymmetry of the Kondo conductance peaks can be traced back to the different spin species which we want to exploit in the next section.

## 8.2. Spin current measurement

In this Section we show that the FM-QD-SC device can be used for spin current measurements. We take advantage of the above observation that AR can be neglected as far as conductance is concerned and obtain the CGF for the separate spins as

$$\ln \chi_{\mathrm{F-QD-S}\sigma} = \ln \chi_{\mathrm{res}\sigma} + \ln \chi_{g\sigma}. \tag{8.5}$$

$\ln \chi_{\mathrm{res}\sigma}$ is obtained from $\chi_{\mathrm{res}}$ by setting $T_{\mathrm{eK}-\sigma} = 0$ and $T_{\mathrm{AKF}} = 0$. Likewise $\ln \chi_{g\sigma}$ is obtained from $\ln \chi_g$ by setting $\ln \chi_{g\sigma} = 1/2 \ln \chi_g$ since the background is assumed to be spin-symmetric. Now we calculate the conductance $G_\sigma$ for the two spin-species as usual from the respective CGF in Eq. (8.5). This allows us to derive the quality factor for spin-filtering in our device

$$q = \left| \frac{G_\uparrow - G_\downarrow}{G_\uparrow + G_\downarrow} \right|, \tag{8.6}$$

along the lines of Dahlhaus et al. [ 2010]. The result is given in Fig. 8.2 using the parameters $T = 0.13\Delta$, $\Gamma = 0.036$, $\Gamma_K = 0.29\Delta$, $P = 0.46$, $\delta_{\mathrm{split}} = 0.16\Delta$ and $T_g = 0.035$ as determinded from the experimental data above. As before we choose for the SC gap in Eq. (8.4) $\tilde{\Delta} = 0.23\Delta$. The quality factor reaches about 70% for voltages around $0.3\Delta$, where the conductance for the majority spin (spin-$\uparrow$) is dominant. For $V/\Delta \approx -0.15$ the minority spin-component (spin-$\downarrow$) is dominant. For $V/\Delta \approx -0.3$ both spin-directions have roughly the same transmission probability ($q$ goes to zero) and for even lower voltages the spin-$\uparrow$ component again takes over, which causes another dip in the $q$ plot in Fig. 8.2 . A possible quality factor of 70% is much better than with a simple FM tunneling contact as there one could only reach a quality factor equal to the polarisation $P$ meaning $\approx 46$ %.
The behavior shown in Fig. (8.2) can be explained by the interplay of the two Kondo resonance peaks that correspond to the separate spin components of the current. For $V/\Delta \approx 0.3$ electronic transport proceeds mainly through the Kondo singlet for spin-$\uparrow$ which explains the large spin polarisation. For negative bias the spin-$\downarrow$ component becomes dominant. The quality factor, however, does not reach the same height as for spin-$\uparrow$ due to the different DOS for the two spin-species in the FM. For large bias electronic transport proceeds mainly through the spin-symmetric background so that the quality factor of spin-filtering goes to zero. Therefore, the capability for spin measurement is a combined effect of the asymmetric DOS in the FM and the splitting of the Kondo resonances by the exchange field $\delta_{\mathrm{split}}$.
Depending on the voltage bias a specific spin direction may be tuned to contribute to charge transfer due to the splitting of the Kondo peak. The Kondo peak defines an almost perfect interface as it proceeds via a collective state of the QD and the FM lead. This makes the FM-QD-SC setup an ideal spin filter. This is of special importance in the case of the double dot Cooper pair splitters investigated in Chapter 7 , where the final proof of entanglement heavily relies on an effective spin measurement as we will discuss in more detail in Chapter 10 .

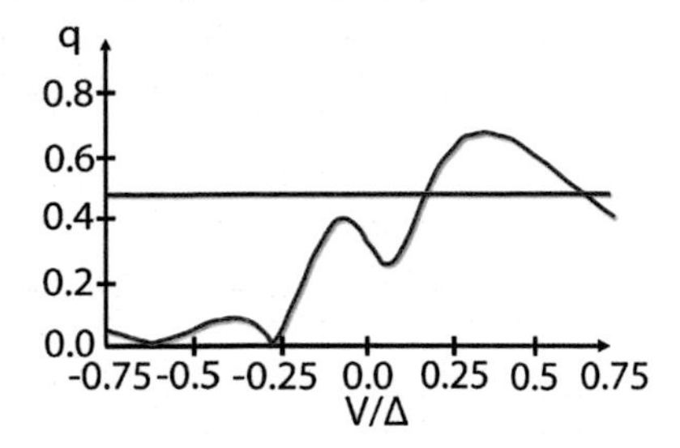


**Figure 8.2.:** Calculation of the quality factor defined in Eq. (8.6) for spin-filtering with the same experimental parameters as used in the fit in Fig. 8.1 . We have neglected ARs since we have shown that they do not considerably change the conductance properties. The quality factor reaches about 70% even at finite temperature as in the experiment. The blue line indicates the quality factor $q = P$ of a simple tunneling junction to a FM with equal polarisation $P = 0.46$.

## 8.3. FM-QD-SC device on resonance

So far we considered the QD in the odd state. Here we turn to the other situation and consider the even charge state of the QD. We did not need to incorporate spin-active scattering in the Kondo regime but with an even population of the QD the Kondo resonance disappears. The absence of the collective state at the Fermi level of the FM allows for the possibility of interface effects. Indeed, we find a pronounced mini-gap feature for an even charge state (Hofstetter et al. [ 2010]) which may be explained by a scenario based on spin-active scattering as in the case of a QPC considered in Section 3.3 [see Fig. 8.3 (a)].

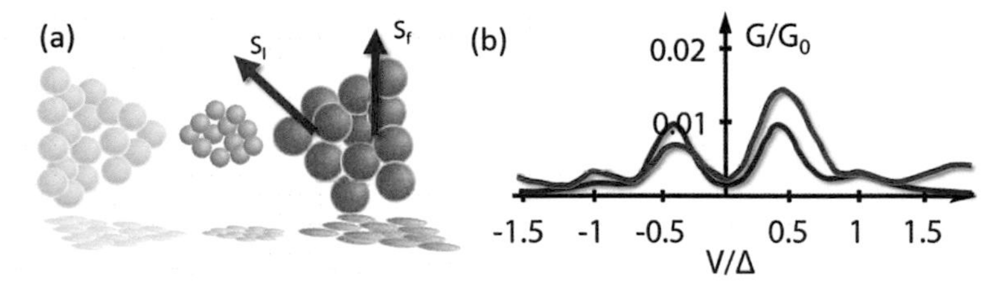


**Figure 8.3.:** **(a)**: The two magnetic moments of the bulk FM $\vec{S}_f$ and the interface $\vec{S}_I$ may be misaligned. This leads to spin-active scattering also in QD junctions.
**(b)**: The conductance for a single channel contact with spin-active scattering is shown as a function of voltage. The theoretical prediction by our model is the blue curve and the red curve refers to the experimental data taken from Hofstetter et al. [ 2010] at a background voltage $V_{BG} = 11.175\,V$. The theoretical fit has been done using the parameters $\Gamma_1 = 0.01\Delta$, $\Gamma_{\text{RS1}} = 0.005\Delta$, $\Gamma_{\text{RS2}} = 0.015\Delta$, $P = 0.46$, $T = 0.1\Delta$, $U = 2\Delta$ and the gap $\tilde{\Delta}$ has been chosen such that the peaks are at the correct position $\tilde{\Delta} = 0.9\Delta$. Furthermore, we can infer $\delta_\uparrow = 0.4\Delta$.

We develop an effective model for an interacting QD in an even charge state. Since the level spacing of the dot ($\delta E \approx 0.4$ meV) is significantly larger than the mini-gap energy, we focus on a single orbital level. The Hamiltonian for a simple resonant level coupled to a FM and a SC with spin-active scattering but still without the Coulomb interaction has already many constituents as in Eq. (6.13) given by

$$H = H_1 + H_S + H_D + H_{\mathrm{TR}}, \tag{8.7}$$
$$H_{TR} = H_{\mathrm{TS}} + H_{\mathrm{TF}1,1} + H_{\mathrm{TF}1,2}. \tag{8.8}$$

$H_D$ has to account for the exchange field $\Delta\epsilon$ induced by the bulk FM. We do so by defining the dot levels as $\epsilon_{D\sigma} = \epsilon_D + \sigma\Delta\epsilon/2$. For simplicity we focus only on a QD on resonance characterised by $\epsilon_D = 0$. The tunnel Hamiltonians in Eq. (8.8) are introduced as in Eq. (6.13) in order to introduce spin-active scattering at the FM-QD interface. We first solve the resonant level case without the exchange field and Coulomb interaction. In the second step we introduce both effects in an effective model.
For computational reasons it is inconvenient to work with the spin-flip tunneling on the dot-FM interface. We choose to rotate the dot fields using Eq. (6.15) and to rewrite Eq. (8.8) in the new basis in order to obtain $H_{\mathrm{TR}} = H_{\mathrm{TD1}} + H_{\mathrm{T1}} + H_{\mathrm{T2}}$, as in Eqs. (6.3) and (6.16). Consequently, the spin-flip tunneling is effectively shifted to the dot-SC interface. Obviously, the above transformation does not change the dot Hamiltonian in Eq. (6.5).
First, we consider the QD without the exchange field ($\epsilon_{D\sigma} = 0$) and onsite Coulomb interaction. In order to access the CGF, we need to introduce two counting fields for the separate leads so that

$$\chi_{RSFa}(\lambda) = \left\langle T_C \exp\left[-i\int_C (T_R^{\lambda(t)} + T_{R2}^{\lambda(t)} + T_{R3}^{\lambda(t)})dt\right]\right\rangle \tag{8.9}$$

where $T_R^{\lambda(t)}$ and $T_{R2}^{\lambda(t)}$ represent $H_{\mathrm{T1}}$ and $H_{\mathrm{T2}}$ with the substitution $c_\sigma(x=0) \rightarrow c_\sigma(x=0)e^{-i\lambda_S(t)/2}$ and $T_{R3}^{\lambda(t)}$ can be obtained from $H_{\mathrm{TD1}}$ with the substitution $\Psi_{1\sigma}(x=0) \rightarrow \Psi_{1\sigma}(x=0)e^{-i\lambda_1(t)/2}$. Using the Hamiltonian approach as before we arrive at the CGF given in Appendix B.3 , Eq. (B.13). The emerging formula is formally identical to the result for the SC-FM-QPC with spin-active scattering in Eq. (3.26) but with energy-dependent transmission coefficients. Above the gap we observe single-electron transmission and spin-flip transmission while below the gap we obtain AR and SAR. The spin-active scattering leads to triplet correlations in the FM. This kind of proximity phenomenon is mediated by the QD instead of a tunneling contact as in Section 3.3 . The new feature in our setup is that the triplet correlations feel the exchange field of the bare FM which allows for a qualitatively new mini-gap feature.
Let us now turn to the situation of an interacting QD with a level splitting given by finite $\Delta\epsilon$. The inclusion of the Coulomb interaction is done in the same way as in Vecino et al. [ 2003] for a Josephson junction: for the AR transmission coefficients the second spin level has to account for the local exchange field $U$ given by Eq. (6.21). This procedure is formally equivalent to a mean field solution including Coulomb interaction. The later analysis of the experimental data shows $\Delta\epsilon \lesssim \Delta$, $\Delta$ being much larger than the tunnel rates of the QD to the FM/SC lead. Using this assumption the result may be greatly simplified: for energies above the gap spin-flip transmissions (and thus $H_{\mathrm{TR2}}$) can be neglected since they involve both spin species. Below the gap AR involves both spin species and can therefore be neglected as well. Therefore we are left with single-electron transmission and SAR and the CGF for the resulting effective model for a QD in the even state is

given by

$$\ln \chi_{\text{es}}(\lambda,\tau) =$$
$$2\tau \int \frac{d\omega}{2\pi} \left( \ln\{1 + \sum_\sigma T_{\text{ese}\sigma}[n_{F+}(1-n_S)(e^{i\lambda}-1) + n_S(1-n_{F+})(e^{-i\lambda}-1)]\}\theta\left(\frac{|\omega|-\Delta}{\Delta}\right) \right.$$
$$\left. + \frac{1}{2}\ln\{1 + \sum_\sigma T_{\text{esAT}\sigma}[n_{F+}(1-n_{F-})(e^{2i\lambda}-1) + n_{F-}(1-n_{F+})(e^{-2i\lambda}-1)]\}\theta\left(\frac{\Delta-|\omega|}{\Delta}\right)\right), \tag{8.10}$$

with the transmission coefficients

$$\begin{aligned} T_{\text{ese}\sigma} &= \frac{4\Gamma_{1\sigma}\Gamma_{S11}}{(\Gamma_{1\sigma}+\Gamma_{S11})^2 + (\omega-\epsilon_{D\sigma})^2}, \\ T_{\text{esAT}\sigma} &= (4\Gamma_{S23}^2\Gamma_{1\sigma}) \,/\, \left[(\omega-\epsilon_{D\sigma})^2(\omega-\epsilon_{D\sigma}+U)^2 + (\Gamma_{S23}^2+\Gamma_{1\sigma}^2)^2 \right. \\ &\quad \left. +\Gamma_{S23}^2(\omega-\epsilon_{D\sigma})(\omega-\epsilon_{D\sigma}+U) + \Gamma_{1\sigma}^2(\omega-\epsilon_{D\sigma})(\omega-\epsilon_{D\sigma}+U)\right], \end{aligned}$$

where we used the shorthand notations $\Gamma_{1\sigma} = \Gamma_1(1+\sigma P)$, $\Gamma_{S11} = \Gamma_{\text{RS1}}|\omega|/\sqrt{|\omega^2-\Delta^2|}$ and $\Gamma_{S23} = 2\sqrt{\Gamma_{\text{RS1}}\Gamma_{\text{RS2}}}\Delta/\sqrt{|\Delta^2-\omega^2|}$ that involve $\Gamma_{\text{RS1}} = \pi\rho_{0S}\gamma_{S1}^2/2$ and $\Gamma_{\text{RS2}} = \pi\rho_{0S}\gamma_{S2}^2/2$. We compare the results of our model described above to the experimental data in Fig. 8.3 (b).
We see that spin-active scattering in the presence of Coulomb interaction may lead to a significant mini-gap feature with a width of $\approx \Delta$ and conductance peaks even higher than the ones associated with the SC gap. The effective model correctly predicts the four-peak structure referring to the SC DOS and the exchange field as one observes in the experiment. It also explains the relation of the mini-gap feature to the FM exchange field: SAR occurs via just a single spin level on the QD since it is associated with the triplet correlation functions in the bare FM. In the presence of the FM exchange field the two spin levels of the QD split. The exchange field therefore causes a splitting of the two SAR conductance peaks which is directly observable as the new mini-gap feature. This splitting is directly given by $\Delta\epsilon$. We do not correctly obtain the asymmetry inside the gap. This is due to the shortcomings of our model. An obvious improvement would be a more sophisticated (including correlation effects) calculation in the interaction as the one in Section 6.3 .
Concerning the approximations we made to arrive at Eq. (8.10) we should note that indeed the exchange field $\Delta\epsilon \approx \Delta \gg \Gamma_{1,\text{RS1},\text{RS2}}$ so that our approximations are justified.
In the experiment the dependence of the subgap feature on an external magnetic field has been investigated as well. One observes that the subgap feature hardly evolves in the magnetic field as long as the SC gap is not fully closed. If the gap closes, also the mini-gap feature gets strongly suppressed. We can analyse the evolution in magnetic field as well, using the model derived above. We add the term $\sigma g\mu_B B/2$ to the positions of the split levels in Eq. (8.10), where $B$ refers to the external magnetic field. We use a typical value (Csonka et al. [ 2008]) of$g = 8$.[1] The evolution of the SC gap is assumed to be given by

$$\Delta(B) = \Delta\sqrt{1-\left(\frac{B}{B_c}\right)^2},$$

and $\Delta$ is substituted in Eq. (8.10) with $\Delta(B)$. For $\Delta(B) = 0$ the effective model given by Eq. (8.10) should not be applicable anymore but since we are only interested in the evolution below the gap we assume the conductance for $\Delta(B) = 0$ to be constant for the voltage range considered here. The result is shown in Fig. 8.4 and compared to experimental data. We find that our

[1] For the charge state analysed in Section 8.1 we obtained$g = 12$. The $g$-factors depend on the charge state considered and we use $g = 8$ here.

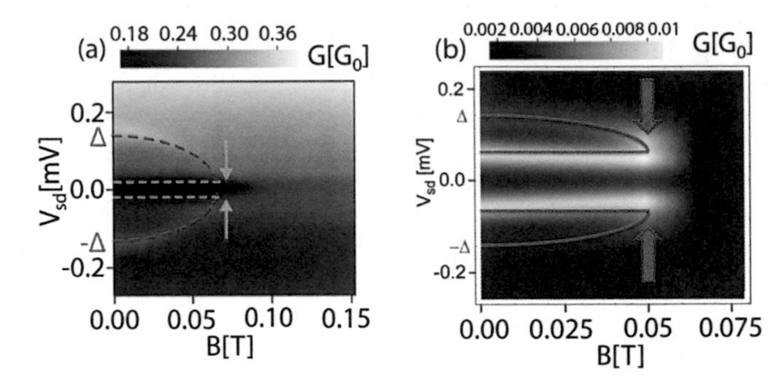


**Figure 8.4.:** The evolution of the subgap feature in a magnetic field: **(a)** experimental data for a typical sample different from the one analysed in Fig. 8.3 (b), (different charge state). One observes a subgap feature at the energy scale of the exchange field indicated in yellow. The feature is suppressed above the critical field of the SC (indicated by yellow arrows).
**(b)**: The plot shows the conductance as a function of magnetic field given by Eq. (8.10) using the same parameters as for Fig. 8.3 (b). We assume$g = 8$ and for vanishing SC gap we assume the conductance to be constant for the voltage range considered here. The critical magnetic field is taken to be $B_c = 64$ mT and $\Delta = 0.14$ meV.

model correctly predicts the qualitative behavior of the mini-gap feature. The gap closes whereas the mini-gap stays in place as long as the gap is not vanishing. This is related to the very large exchange field observed in the experiment. $\Delta\epsilon = 0.8\Delta$ corresponds to a critical magnetic field of the subgap feature of $B_{c,\text{subgap}} = 0.19$ T, which is much larger than the critical magnetic field of the SC. Therefore, we conclude that our model delivers a qualitatively correct description of the underlying physics. According to our explanation the splitting of the SAR peaks due to the exchange field gives direct evidence for the triplet correlations due to spin-active scattering since normal (spin-symmetric) AR conductance peaks cannot split up in an applied magnetic field. In this way the spin-active scattering can be identified in simple transport experiments in a way similar to the explicit investigation of Andreev bound states (Hübler et al. [ 2012]).

## 8.4. Conclusions

In conclusion, we have investigated SC-QD-FM hybrids realized using an InAs nanowire both in the even and odd charge state. Even though an approach as in Section 6.3 could in principle be followed the numerous energy scales present in the problem make straightforward progress cumbersome. Therefore, we used an effective model to investigate the behavior of the odd charge state in the Kondo limit. In this case the Kondo effect imposes a strong asymmetry between the coupling of the SC and the FM to the QD. Spin-active scattering at the interface is strongly suppressed making the device an ideal tool for spin measurements in Cooper pair splitters as we will discuss in more detail in Chapter 10 . Spin-active scattering turns out to be 'switched on' in an even charge state of the QD. There it induces triplet correlations that lead to an observable mini-gap feature. Furthermore, our model allows to reproduce and interpret the evolution of the mini-gap feature in an external magnetic field.

# Chapter 9

## Normal-quantum dot-superconductor junctions

In Section 5.2 we briefly introduced the Anderson-Holstein model, describing the coupling of a QD level to a phonon-mode. Especially when concerned with QDs formed by molecules (Liang et al. [2002]) or suspended carbon nanotubes (Leturcq et al. [ 2009]) the charging of the QD can lead to a substantial deformation of the QD itself. This leads to a coupling of the electronic to the vibrational degrees of freedom. For suspended carbon nanotubes this coupling is large and leads to the pronounced steps in the $I - V$ characteristics as in Fig. 5.8 . This strongly non-linear behavior allows to use these junctions as transistors. The problem is quite difficult so that in most cases one hast to resort to numerical approaches even for normal terminals (Albrecht et al. [ 2012b], Mühlbacher and Rabani [ 2008]).
Therefore, approaching the full complexity of the Anderson-Holstein model with a SC lead seems out of question and this Chapter will only provide first steps towards a full understanding. In order to verify a diagrammatic Monte Carlo (diagMC) approach to the problem which was developed alongside the analytical calculations that will be presented here we first discuss the transient behavior of the resonant level model with a SC lead and compare it to the diagMC results. Such a comparison allows us to conclude that diagMC provides converged results for the current. We will discuss the relevance of our results for quantum dots in the deep Kondo limit using the effective description of Section 6.4 . Then, we describe coupling of the QD to a phonon mode using a rate equation approach and again compare the results to diagMC data. We follow the argumentation in Albrecht et al. [ 2012a].

## 9.1. Model

A sketch of the model considered here is shown in Fig. 9.1 .
The full Hamiltonian of the Anderson-Holstein model with a SC lead consists of seven contributions

$$H_{\text{sys}} = H_S + H_1 + H_D + H_{\text{TD}} + H_U + H_{\text{Ph}} + H^{(I)}_{\text{D,Ph}}. \tag{9.1}$$

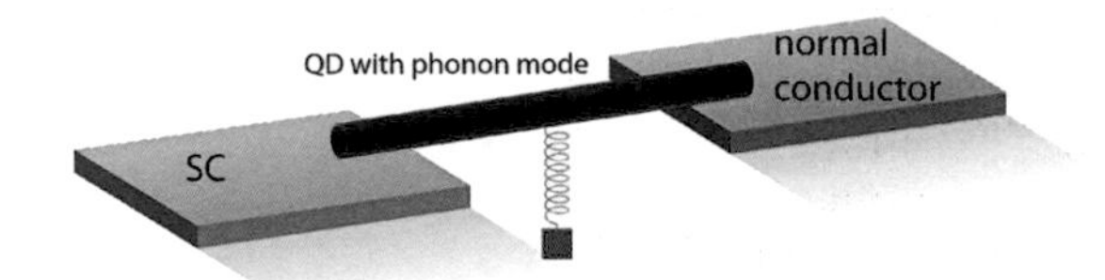


**Figure 9.1.:** Sketch of the system: a QD with strong onsite Coulomb interaction is assumed to be in the Kondo regime and its resonance width is given by $\Gamma_K$. It is coupled to a normal conductor and a SC. The QD is subjected to a finite voltage bias and coupled to a phonon mode.

The first five terms have been introduced before in Eqs. (2.7), (3.1), (6.5), (6.3) and (6.21). The novel ingredient is the electron-phonon interaction

$$H^{(I)}_{\text{D,Ph}} = \sum_{\sigma} d^+_\sigma d_\sigma M_0(b^+ + b), \tag{9.2}$$

which couples the electronic degrees of freedom on the dot to a local phonon mode described by

$$H_{\text{Ph}} = \omega_0(b^+ b + 1/2). \tag{9.3}$$

In the following we do not treat the onsite Coulomb repulsion explicitely and set $H_U = 0$ in Eq. (9.1). We will discuss the conditions under which our results are also relevant for the case of finite $U$.

## 9.2. Analytical approach to transient behavior

The diagMC starts with a system in a 'preparative' non-equilibrium and evolves it towards the steady state. In order to verify the convergence of the diagMC results, we start with an analytical description of the time-dependent behavior of the model introduced above assuming an initially uncoupled dot and describing the approach to the steady state current for an instantaneous switch-on of the tunnel couplings.
For the transient behavior we only consider the case $M_0 = 0$, meaning without electron-phonon interaction. Our choice of parameters should allow to make contact to the case of strong onsite interaction and an odd number of electrons on the QD. In this case we can use the effective model described in Section 6.4 for small$T_K/\Delta$. Our study in Section 8.1 revealed that in new experiments on the Kondo resonance in SC hybrids the background DOS can be very small compared to the Kondo peak so that the system may indeed be effectively described by a resonant level between the SC and the normal conductor with strongly asymmetric hybridisations of the QD with the SC and the normal lead for voltages not much larger than $T_K$. The Kondo resonance is then pinned to the Fermi level in the normal lead so that $\epsilon_D = \mu_1$. The case of strongly asymmetric hybridisations of the QD with the leads directly allows for further simplification of the problem: below the gap electronic transport in SC hybrids is due to AR but our investigation in Sections 6.4 and 8.1 revealed that AR is strongly suppressed due to the strongly asymmetric hybridisation and can be neglected as far as the current is concerned. Therefore, we can treat the SC as a normal conductor with the energy-dependent DOS given by $\rho_S(\omega)$. This corresponds to the semiconductor model introduced in Section 3.1 .

Nonetheless, one could argue that such mapping is not adequate since the time scale for the central Kondo peak to fully develop should be given by $1/T_K$ (Nordlander et al. [1999]). Recent studies of the transient behavior of the Anderson impurity model, however, revealed that the central peak develops much faster and only the Hubbard subbands at $\epsilon = \pm U/2$ develop on a slow time scale (Mühlbacher et al. [2011]). Therefore, the model considered should not only be a specific case of the resonant level model but should be at least a good approximate description of the behavior in the deep Kondo limit.
Therefore, we treat the resonant level model without ARs and strongly asymmetric hybridisations of the leads including the SC DOS. Expectation values separate into spin-$\uparrow$ and spin-$\downarrow$ contributions so that we work with spinless operators from now on. As in Schmidt et al. [2008] we have to deal with two different situations: (i) the dot level is empty $n_0 = 0$ and (ii) the dot is populated by one electron $n_0 = 1$. Due to the simple structure of $H_D$ the time evolution of the decoupled dot is trivial and we obtain the following GF

$$\mathbf{D}_0(t) = e^{-i\epsilon_D t}\begin{bmatrix} -i[\theta(t)(1-n_0) - \theta(-t)n_0] & in_0 \\ -i(1-n_0) & -i[\theta(-t)(1-n_0) - \theta(t)n_0] \end{bmatrix}.$$

For the retarded and advanced components we have

$$D_0^R(t) \;=\; D_0^{--}(t) - D_0^{-+}(t) = -i\theta(t)e^{-i\epsilon_D t}, \;\; D_0^A(t) = D_0^{-+}(t) - D_0^{++}(t) = i\theta(-t)e^{-i\epsilon_D t}.$$

Compared to Schmidt et al. [2008] we now have to deal with two different lead GFs. In both cases the retarded GF has the form

$$g_\alpha^R(t) = -i\theta(t)\int d\omega \rho_\alpha(\omega)e^{-i\omega t}, \tag{9.4}$$

where $\alpha = 1, S$ and $\rho_S(\omega)$ depends on energy due to the SC correlations but also $\rho_1(\omega)$ could become energy-dependent due to a finite bandwidth as discussed in Schmidt et al. [2008]. For the diagMC a soft cutoff of the form

$$\rho_1(\omega) \;=\; \frac{\rho_{01}}{[1+e^{\beta(\omega-\epsilon_c)}][1+e^{-\beta(\omega-\epsilon_c)}]}, \tag{9.5}$$

for the normal terminal is used, while for the SC lead a hard cutoff is taken so that

$$\tilde{\rho}_S(\omega) \;=\; \rho_S(\omega)\theta(\epsilon_c - |\omega|). \tag{9.6}$$

However, for the analytical approach here we consider the case of a wide flat band $\rho_1(\omega) = \rho_{01}$ and $\rho_S(\omega) = \rho_{0S}|\omega|/\sqrt{\omega^2 - \Delta^2}$. One obtains the full Keldysh matrix

$$g_\alpha(\omega) = i2\pi\rho_\alpha(\omega)\begin{bmatrix} n_\alpha - \frac{1}{2} & n_\alpha \\ -(1-n_\alpha) & n_\alpha - \frac{1}{2} \end{bmatrix}. \tag{9.7}$$

The retarded and advanced components in the wide band limit are given by $g_\alpha^R(\omega) = -i\pi\rho_\alpha(\omega)$ and $g_\alpha^A(\omega) = [g_\alpha^R(\omega)]^*$. The GFs of the coupled system can now be found for any time dependence of $\gamma_\alpha(t)$. Here we concentrate on the case of sudden switching where $\gamma_\alpha(t) = \gamma_\alpha\theta(t)$. The time evolution of the retarded GF is given by the standard expression (Caroli et al. [1971], Jauho et al. [1994])

$$D^R(t,t') = D_0^R(t-t') + \int_0^\infty dt_2 K(t,t_2)D^R(t_2,t'), \tag{9.8}$$

where

$$K(t,t_2) \;=\; K_S(t,t_2) + K_1(t,t_2) \tag{9.9}$$

$$\;=\; \int_0^\infty dt_1 D_0^R(t-t_1)\Sigma_1^R(t_1-t_2) + \int_0^\infty dt_1 D_0^R(t-t_1)\Sigma_S^R(t_1-t_2), \tag{9.10}$$

is the kernel involving the lead self-energies $\Sigma_\alpha^R(t) = \gamma_\alpha^2 g_\alpha^R(t)$. The lead self-energy $\Sigma = \Sigma_S + \Sigma_1$ is therefore given by two parts referring to the SC and the normal lead

$$\Sigma_S^R(t) = -\frac{i\theta(t)}{4\pi}\int_{-\infty}^{\infty} d\omega \frac{e^{-i\omega t}\Gamma_{\rm ST}|\omega|}{\sqrt{\omega^2-\Delta^2}}\theta\left(\frac{|\omega|-\Delta}{\Delta}\right), \tag{9.11}$$

$$\Sigma_1^R(t) = -\frac{i\theta(t)}{4\pi}\int_{-\infty}^{\infty} d\omega\, e^{-i\omega t}\Gamma_{\rm 1T}, \tag{9.12}$$

with $\Gamma_{\rm ST} = 2\pi\rho_{0S}\gamma_S^2$ and $\Gamma_{\rm 1T} = 2\pi\rho_{01}\gamma_1^2$.
From the calculation for the normal conducting case we know that the integral equation for the retarded GF involving the normal lead

$$D_1^R(t,t') = D_0^R(t-t') + \int_0^\infty dt_2 K_1(t,t_2) D_1^R(t_2,t'), \tag{9.13}$$

can be solved by iterations (Jauho et al. [1994]) leading to

$$D_1^R(t-t') = -i\theta(t-t')e^{-i\epsilon_D(t-t')}e^{-\Gamma_{\rm 1T}(t-t')/2}. \tag{9.14}$$

For the SC part we first do the Fourier transformation in Eq. (9.11) following Eckern et al. [1984]

$$\int_0^\infty d\epsilon \frac{\epsilon}{\sqrt{\epsilon^2-\Delta^2}}e^{-i\epsilon t}\theta\left(\frac{\epsilon-\Delta}{\Delta}\right) = K_1(\Delta t), \tag{9.15}$$

where $K_1$ is the modified Bessel function. Using this result we arrive at

$$K_S(t,t') = -\frac{\theta(t-t')}{\pi}\int_{t'}^{t} dt_1 e^{-i\epsilon_D(t-t_1)}\, K_1[\Delta(t_1-t')],$$

which we have to evaluate numerically in the expression for the full retarded GF

$$D^R(t,t') = D_1^R(t-t') + \int_0^\infty dt_2 K_S(t,t_2) D^R(t_2,t'), \tag{9.16}$$

that involves already all terms in the tunnel coupling to the normal lead. Due to the numerical evaluation of $K_S(t,t')$ we cannot solve this Dyson equation exactly. However, from our considerations above we know that $\Gamma_{\rm ST} \ll \Gamma_{\rm 1T}$. In this case it suffices to iterate Eq. (9.16) two times.
A simply accessible quantity to calculate is the time-dependent dot population $n(t) = \langle d^+(t)d(t)\rangle$ which can be directly expressed via the off-diagonal Keldysh GF

$$n(t) = -iD^{-+}(t,t). \tag{9.17}$$

The '$-+$'-component of the GF can be expressed via the already known retarded GF using (Langreth [1976])

$$D^{-+} = (1 + D^R\Sigma^R)D_0^{-+}(1+\Sigma^A D^A) + D^R\Sigma^{-+}D^A, \tag{9.18}$$

where in the product notation integrations over omitted time variables are implied. This relation is especially suited for the case of an initially empty dot since then $D_0^{-+} = 0$ and only the last term contributes. In the following we consider the case of the dot being initially empty.
We know from Sections 6.4 and 8.1 that typical experiments have $T_K = \Gamma_{\rm 1T} \lesssim \Delta$. For simplicity we consider $\Gamma_{\rm 1T} = \Delta$ and $\Gamma_{\rm ST} = 0.05\Delta$ in order to stay within the limits of applicability of our approach. Typical results for the time-dependent dot population are shown in Fig. 9.2 (a).
In the normal conducting case the dot population may shoot over its asymptotic steady state value

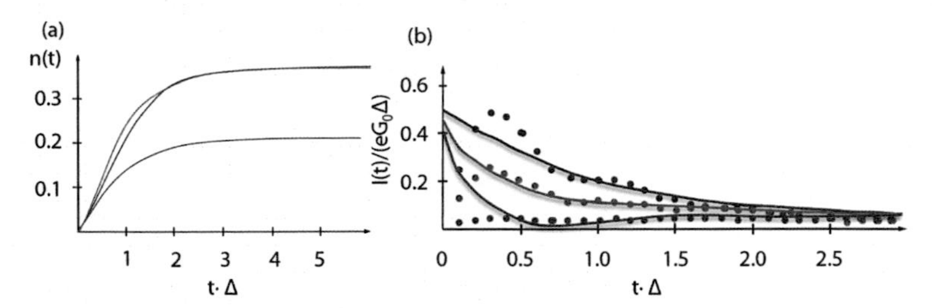


**Figure 9.2.: (a)**: Time-dependent dot population $n(t)$. The graph shows the result for $T = 0.1\Delta$, $\Gamma_{1T} = \Delta$, $\Gamma_{ST} = 0.05\Delta$ and $\epsilon_D = -V$. The lowest (red) curve refers to $V = 0$, the middle (blue) curve to $V = \Delta$ and the upper (green) curve is for $V = 2\Delta$.
**(b)**: Time-dependent current through the QD. We choose the parameters $V = 2\Delta, \epsilon_D = -2\Delta$, $\Gamma_{1T} = \Delta$, $\Gamma_{ST} = 0.05\Delta$ and $T = 1/\beta = 0.1\Delta$. The upper (blue) dots show the result via the diagMC method for the right current and the lower (red) dots correspond to the result for the left current. The blue and red curves corresponds to the analytical results. The middle (green) curve is the average current from Eq. (9.23).

and relax to the steady state accompanied by a number of oscillations (Schmidt et al. [ 2008]). In the SC case treated here we observe no such overshooting and also population oscillations are not observed due to the strongly asymmetric couplings of the dot to the leads.
The more important observable is the current through the system. We can rewrite the current on the normal conducting side in terms of Keldysh GFs and perform the Keldysh disentanglement as in Schmidt et al. [ 2008]

$$I_R(t) = I'_R(t) + I''_R(t), \tag{9.19}$$

$$I'_R(t) = -\gamma_1^2 \mathrm{Re} \int_0^\infty dt_1 g_1^R(t,t_1) D^K(t_1,t), \quad I''_R(t) = \gamma_1^2 \mathrm{Re} \int_0^\infty dt_1 D^R(t,t_1) g_1^K(t_1,t). \tag{9.20}$$

The first contribution is essentially given by the time-dependent dot population

$$I'_R(t) = \frac{\Gamma_{1T}}{2}\theta(t)[1-2n(t)]. \tag{9.21}$$

Therefore, as in the case of normal contacts, there is an instantaneous current onset $I_R(0) = \Gamma_{1T}/2$ which stems from the fact that we assume a sudden switching and have taken the wide band limit so that electrons at arbitrarily high energies are able to occupy the dot in correspondingly fast processes.
Access to the current for the contact between the SC and the QD can either be obtained via the GFs directly but this way is cumbersome since the calculation of the current for the contact of the QD to the normal conductor is easier due to the simpler structure of the GFs. Instead, we can use the displacement current $I_{\text{disp}}$ which is given by the time derivative of $n$ (Büttiker et al. [ 1993], Fransson et al. [ 2002])$I_{\text{disp}}(t) = \frac{dn}{dt}$. It represents the difference in currents between the right and left side

$$I_{\text{disp}}(t) = I_R(t) - I_L(t) \Leftrightarrow I_L(t) = I_R(t) - I_{\text{disp}}(t). \tag{9.22}$$

Consequently $I_L(t)$ will show a similar current onset as $I_R$ since calculating $D^R$ in Eq. (9.16) iterated two times in $\Gamma_{ST}$ regularizes the initial derivative of $n(t)$. We may also define the average current through the junction as

$$I(t) = [I_R(t) + I_L(t)]/2. \tag{9.23}$$

We compare the result from the analytical approach choosing $\Gamma_{1T} = \Delta$, $\Gamma_{ST} = 0.05\Delta$ and $\beta = 10\Delta$ to the diagMC result in Fig. 9.2 (b).
We observe good agreement of both methods. We should stress again that the analytical calculation has been performed in the wide flat band limit and iterated two times in $\Gamma_{ST}$, while the diagMC approach uses a finite cutoff which leads to the slight discrepancies for small times. For larger times the results still agree very well but simulations for larger time scales become increasingly time-consuming. In accordance with previous studies for normal contacts (Mühlbacher and Rabani [2008], Schmidt et al. [ 2008]) we observe a fast approach to the steady state even in the situation of strongly asymmetric couplings. Therefore, we conclude that the diagMC scheme, also in the presence of the diverging SC DOS, provides converged results despite the notorious fermionic and dynamical sign problems. The approach to the steady state does not proceed on a long time scale given by $1/\Gamma_{ST}$ but on the short time scale $1/\Gamma_{1T}$ which allows for a reliable and fast simulation of the system. We do not see pronounced effects of the SC DOS due to the small hybridisation of the SC with the QD.

## 9.3. Rate equation approach to electron-phonon coupling

The next step is to include the electron-phonon coupling in Eq. (9.1). We treat the same situation as in Section 9.2 : the SC is described like a normal metal with the DOS$\rho_S(\omega)$ and the hybridisations of the SC and the normal metal are strongly asymmetric. In this situation we have a resonant level model coupled to phonons as in Kast et al. [ 2011], however, with a SC DOS.
For the resonant level model it is convenient to work with dressed electronic states by applying a polaron transformation $U_p = \exp[(M_0/\omega_0)d^+d(b^+ - b)]$ which leads to a Hamiltonian where the electron-phonon interaction $H^{(I)}_{\mathrm{D,Ph}}$ is completely absorbed in the tunnel part of the Hamiltonian meaning that we are left with Eq. (9.1) in the form of $H_{\mathrm{sys}} = H_L + H_R + \tilde{H}_{\mathrm{TD}} + H_{\mathrm{Ph}} + \tilde{H}_D$, where

$$\begin{aligned} \tilde{H}_{\mathrm{TD}} &= \gamma_1[\Psi^+_{k1}(x=0)e^{(M_0/\omega_0)(b^+-b)}d + \mathrm{H.c.}] + \gamma_S[c^+_k(x=0)e^{(M_0/\omega_0)(b^+-b)}d + \mathrm{H.c.}], &(9.24)\\ \tilde{H}_D &= (\epsilon_D + M_0^2/\omega_0)(b^+b + 1/2). &(9.25)\end{aligned}$$

We absorb the polaron shift of the dot energy by a redefinition of $\epsilon_D \to \epsilon_D - M_0^2/\omega_0$. For the rate equation we assume that the GFs for the dot including the phonon interaction factorise into an electronic and a phonon part where the latter can be fully characterised by the phonon correlation function (Kast et al. [ 2011])

$$e^{J(t)} = \left\langle e^{(M_0/\omega_0)[b^+(0)-b(0)]}e^{-(M_0/\omega_0)[b^+(t)-b(t)]}\right\rangle. \qquad (9.26)$$

Again using the self energies introduced in Eqs. (9.11) and (9.12) we obtain the forward and backward rates onto the dot ($\Gamma_{R1}$, $\Gamma'_{R2}$) and away from the dot ($\Gamma'_{R1}$, $\Gamma_{R2}$) following Kast et al. [2011]

$$\begin{aligned} \Gamma_{R1}(V) &= \int \frac{d\omega}{2\pi}\frac{\Gamma_{ST}|\omega|}{\sqrt{\omega^2-\Delta^2}}\theta\left(\frac{|\omega|-\Delta}{\Delta}\right)n_S P(\omega - V),\ \ \Gamma_{R2}(V) = \int \frac{d\omega}{2\pi}\Gamma_{1T}(1-n_1)P(\omega+V),\\ \Gamma'_{R1}(V) &= \int \frac{d\omega}{2\pi}\frac{\Gamma_{ST}|\omega|}{\sqrt{\omega^2-\Delta^2}}\theta\left(\frac{|\omega|-\Delta}{\Delta}\right)(1-n_S)P(\omega + V),\ \ \Gamma'_{R2}(V) = \int \frac{d\omega}{2\pi}\Gamma_{1T}n_1 P(\omega-V).\end{aligned}$$

The inelastic tunneling processes associated with the energy emission and absorption of phonons are described by the function $P(\omega)$ being the Fourier transform of $e^{J(t)}$ introduced in Eq. (9.26). For this correlation function we assume that the phonons are thermally distributed, which may be due to coupling to a thermal environment given by the substrate or a backgate. The effect of coupling to an external bath can be characterised by an additional coupling constant $\gamma_B$ and in the following we assume the bath to be purely ohmic. In this case the phonon spectral density has Lorentzian shape (Leggett et al. [ 1987])

$$J_B(\omega) \;=\; \frac{\gamma_B \omega}{[(\omega/\omega_0)^2 - 1]^2 + [\gamma_B \omega_0 \omega/(2M_0^2)]^2}. \tag{9.27}$$

The phonon correlation function can now be calculated analytically (Kast et al. [ 2011]) leading to

$$P(\omega) = \frac{e^{-\rho_{\gamma_B}}}{\pi}\mathrm{Re}\left\{\sum_{k,l=0}^{\infty} \frac{\rho_{\gamma_B,a}^k}{k!}\frac{\rho_{\gamma_B,e}^l}{l!}\frac{i}{\omega + \Omega_0 k - \Omega_0^* l + i\Gamma_{\mathrm{tot}}/2}\right\}, \tag{9.28}$$

where $\Omega_0 = \omega_0\xi + i\gamma_B/2$, $\xi = \sqrt{1 - \gamma_B^2/(4\omega_0^2)}$ and $\Gamma_{\mathrm{tot}} = \Gamma_{1\mathrm{T}} + \Gamma_{\mathrm{ST}}$. The functions $\rho_{\gamma_B}$, $\rho_{\gamma_B,\alpha}$ and $\rho_{\gamma_B,e}$ are given by

$$\rho_{\gamma_B} \;=\; \frac{M_0^2}{2\omega_0\sqrt{\omega_0^2 - \gamma^2/4}}\left[\frac{\coth\left(\frac{\beta\Omega_0}{2}\right)}{\Omega_0^2} + \frac{\coth\left(\frac{\beta\Omega_0^*}{2}\right)}{(\Omega_0^*)^2}\right], \tag{9.29}$$

$$\rho_{\gamma_B,a} \;=\; \frac{M_0^2}{2\omega_0\sqrt{\omega_0^2 - \gamma_B^2/4}}\frac{\coth\left(\frac{\beta\Omega_0}{2}\right) - 1}{\Omega_0^2}, \; \rho_{\gamma_B,e} = \frac{M_0^2}{2\omega_0\sqrt{\omega_0^2 - \gamma_B^2/4}}\frac{\coth\left(\frac{\beta\Omega_0^*}{2}\right) - 1}{(\Omega_0^*)^2}. \tag{9.30}$$

Using the above defined rates in the master equation for the dot population we can solve for the steady state and derive the current which is given by

$$I(V) \;=\; \frac{\Gamma_1\Gamma_2 - \Gamma_1'\Gamma_2'}{\Gamma_1 + \Gamma_2 + \Gamma_1' + \Gamma_2'}. \tag{9.31}$$

In Fig. 9.3 we have calculated the current for different values of the coupling to the external bath.

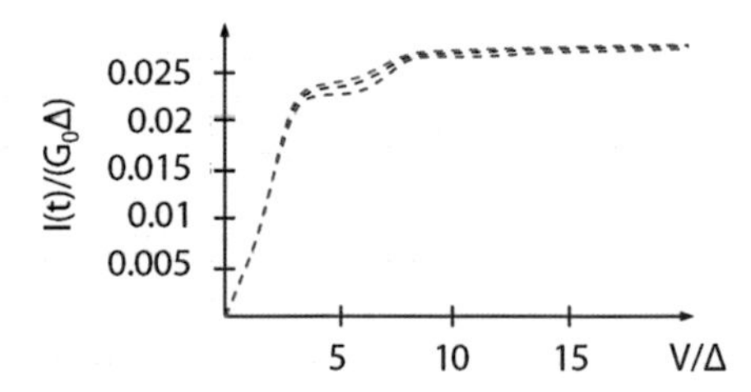


**Figure 9.3.:** Steady state current as a function of voltage given by the rate equation in Eq. (9.31). The result is for $\Gamma_{1\mathrm{T}} = 0.1\Delta$, $\Gamma_{\mathrm{ST}} = 0.9\Delta$ and a moderate coupling $M_0 = 2\Delta$, $\omega_0 = 2\Delta$ for the Einstein phonon and $\beta = 10\Delta$. The lower (red) curve corresponds to $\gamma_B = 0$, the middle (blue) curve corresponds to $\gamma_B = 0.3\Delta$ and the upper (green) curve corresponds to $\gamma_B = 0.5\Delta$.

We observe the typical saturation of the current at high voltages and also slight steps in the current indicating the presence of the phonon modes. As $\gamma_B$ increases the steps are washed out and eventually disappear. A slight coupling to the environment does not change our results very much

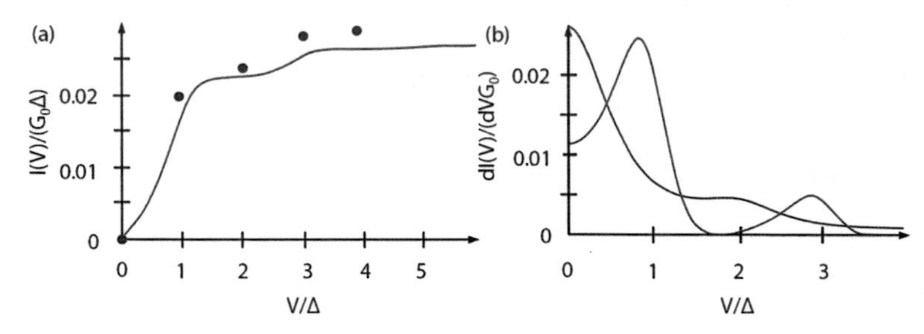


**Figure 9.4.: (a)**: Steady state current as a function of voltage given by the rate equation in Eq. (9.31) (red curve) using the cutoffs as in Eqs. (9.5), (9.6) and given by the diagMC approach (blue dots). The result is for $\Gamma_{1T} = 0.9\Delta$, $\Gamma_{ST} = 0.1\Delta$ and a moderate coupling $M_0 = 2\Delta$, $\omega_0 = 2\Delta$ for the Einstein phonon and $\beta = 10\Delta$.
**(b)**: Conductance $dI(V)/dV$ using the same parameters as for the plot in (a). The comparison of the conductance in the superconducting (red) and normal (blue) case shows that the conductance peaks due to the phonon sidebands become much more pronounced.

so we will treat the case $\gamma_B = 0$ in the following. We see, however, that pronounced phonon steps can be seen at small voltages. Therefore, we want to investigate this range of voltages now more closely using the exact diagMC approach.
We would like to stress that the time-dependent dynamics discussed in Section 9.2 can also be described with the diagMC method in the presence of electron-phonon interaction but the rate equation approach provides no direct access to them. We therefore compare results for the steady state current in the presence of electron-phonon interaction. We use again a typical moderate electron-phonon coupling $M_0$ and compare our results for the $I - V$, see Fig. 9.4 (a). We use the same cutoffs (Eqs. (9.5) and (9.6)) for the rates as for the diagMC.
We observe good agreement in the voltage range considered, so that we can use the rate equation approach in order to investigate specific effects of the SC and phonon DOS. The convolution of these two quantities leads to steps in the current not at multiples of the oscillator frequency but equally spaced by a distance of $V \approx \omega_0$.
These effects appear even more pronounced in the conductance $dI(V)/dV$. Since the rate equation approach provides us with a continuous curve we can calculate the derivative shown in Fig. 9.4 (b) and compare it to the normal conducting case.
We observe that the conductance steps become much more pronounced compared to the normal conducting case due to the SC DOS. There is also an offset of $\Delta$ due to the absence of quasiparticles in the SC at energies below the gap.
The mapping described in Section 9.2 of the model including onsite interaction to a resonant level in the deep Kondo limit is not expected to hold in the presence of electron-phonon interaction. Still, at moderate interaction strength of the phonon mode with the electrons on the QD it has been shown in Paaske and Flensberg [ 2005] that the Kondo effect does not disappear at once but rather persists for moderate electron-phonon interactions, however, associated with a change in $T_K$ and the appearance of phonon sidebands. Since the mapping to an effective Kondo model is still possible (Paaske and Flensberg [ 2005]) a good qualitative description could be obtained from a rate equation approach (König et al. [ 1996]). We can therefore assume that the mapping to the resonant level as described in Section 9.2 still works, however, keeping in mind that$T_K$ will change due to electron-phonon interaction (Paaske and Flensberg [ 2005]).
Consequently, concerning the model described in Section 9.1 we would draw the conclusion that

compared to the normal conducting case analysed in Paaske and Flensberg [2005] the phonon sidebands will be much more pronounced in the presence of a SC lead. However, we again emphasize that our model should only give a qualitatively correct description of the case of a Kondo QD in the presence of electron-phonon interaction.
This shortcoming of our consideration is not due to the diagMC approach. Indeed, it could easily be generalized to obtain a full picture of the Anderson-Holstein model with a SC lead. We are limited by the analytical models that do not allow for an easy generalization and are therefore limited to the study of effective models. However, analytical models can be used to access higher order correlation functions as the noise also for the transient case (Joho et al. [2012]).

## 9.4. Conclusions

In conclusion, we have investigated the Anderson-Holstein model with a SC lead. In the first part (Sections 9.1, 9.2) we have investigated the transient dynamics of the Anderson model with a SC lead using a mapping of the deep Kondo limit to the resonant level model. We have found good agreement between the analytical and a diagMC approach so that we have been able to conclude that we obtain converged results for the steady state current from the diagMC method. In the second part (Section 9.3) we have then used the diagMC approach to obtain the current also in the presence of electron-phonon interaction. We have compared this effective approach to the Anderson-Holstein model with a SC lead to a rate equation approach and observed good agreement. We find that the transient behavior is dominated by a large overshooting of the intial current and a rapid but smooth approach to the steady state current. Electron-phonon interaction leads to more pronounced sidebands compared to the normal conducting case.

Chapter 10

# Entanglement witnessing in SC beamsplitters

In the introduction of Chapter 7 we argued that SCs are an ideal on-chip source of EPR pairs since entanglement is inherent to the spin singlet ground-state of a Cooper pair. In Chapter 7 and 8 we investigated the mechanism of splitting Cooper pairs in detail. However, splitting of Cooper pairs does not automatically mean that we have created a stream of entangled electrons on tap (Merali [2009]). Rather, splitting is only the first step (Schönenberger [2010]) and the verification of entanglement is the second one (see also Chapter 7). Nonetheless, one can be optimistic that solid-state physics, the scientific mainstay of digital computation, will provide a suitable arena for quantum computation using SC circuits. This Chapter will show that present day technique allows to verify the presence of entanglement in SC beamsplitters which represents the first step towards quantum computation (Loss and DiVincenzo [1998]). We will follow the discussion from Soller et al. [2012e].

## 10.1. Entanglement witnessing

The process of CPS allows to split electrons arriving at two distant observers Alice and Bob as indicated in Fig. 10.1. We start with the simplest setting as in Di Lorenzo and Nazarov [2005]. Both Alice and Bob can perform spin measurements in two directions $\mathbf{m}_A, \mathbf{m}'_A$ and $\mathbf{m}_B, \mathbf{m}'_B$, respectively, see Fig. 10.1.
As in optical experiments we define the Bell parameter to be

$$\epsilon = |E(\mathbf{m}_A, \mathbf{m}_B) + E(\mathbf{m}'_A, \mathbf{m}_B) + E(\mathbf{m}_A, \mathbf{m}'_B) - E(\mathbf{m}'_A, \mathbf{m}'_B)|, \quad (10.1)$$

where the correlator is given by $E(\mathbf{m}, \mathbf{m}') = P_{\mathbf{mm}',++} + P_{\mathbf{mm}',--} - P_{\mathbf{mm}',+-} - P_{\mathbf{mm}',-+}$. $P_{\mathbf{mm}',\sigma\sigma'}$ refers to the probability to observe an electron pair in detectors with directions $\mathbf{m}, \mathbf{m}'$ with spin directions $\sigma = \pm$ and $\sigma' = \pm$. In our case the detectors for electrons will be the four FM terminals ($F1$ - $F4$) and the source will be the SC ($S$), see Fig. 10.2 (a). The DOS of a FM is spin-dependent (see Section 3.2), so that the detection of an electron in a FM is also an (imperfect) spin measurement. We write the magnetisation direction $\mathbf{g}_i$, $i = 1, 2, 3, 4$ as $\mathbf{g}_i = \mathbf{m}_i P_i$ with

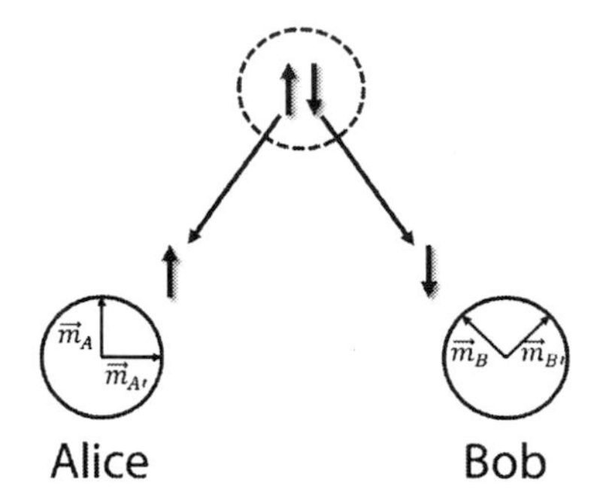


**Figure 10.1.:** A schematic for a Bell measurement for the electrons of a Cooper pair. First, the two electrons with opposite spin are separated to the observers Alice and Bob. Afterwards projection measurements along the axes $\mathbf{m}_{A,A'}$ and $\mathbf{m}_{B,B'}$ are performed.

a unit vector $\mathbf{m}_i$ that indicates the direction. In the following we want to assume that all four FMs attached have the same polarisation and the same coupling to the SC in order to simplify notation. Additionally, in order to have spin detectors the magnetisation direction of $F1$, $F2$ and $F3$, $F4$ is assumed to be pairwise antiparallel as also indicated in Fig. 10.2 (a). We therefore write $\mathbf{g}_1 = -\mathbf{g}_2 = \mathbf{g}_A$, $\mathbf{g}_3 = -\mathbf{g}_4 = \mathbf{g}_B$ referring to Alice and Bob. The case for non-equal polarisations in a special case has been treated in Morten et al. [2008]. It does not lead to a qualitatively different behavior.

In a first approximation for small bias $V \ll \Delta$, let us now forget about the DOS of the SC and focus on processes where a Cooper pair comes from the SC, is split and transferred further to the separate leads. This transfer is enforced by applying the bias voltage $V$ between the leads $F1$ - $F4$ and $S$. The leads are all kept at the same chemical potential. Then we can write the CPS conductance for small bias $V$ and for transfer from $S$ to $Fi$ and $Fj$ as

$$G_{ij} = f(V, \delta r, \Sigma)(1 - \mathbf{g}_i \mathbf{g}_j), \tag{10.2}$$

where $f(V, \delta r, \Sigma)$ is a function of the applied bias $V$, the width $\delta r$ of the SC (SC finger in Fig. 10.2) and a possible interaction described by a self-energy $\Sigma$. For the case of simple tunnel contacts between the SC and the FMs $f(V, \delta r, \Sigma)$ is just a constant $f(V, \delta r, \Sigma) = 16\Gamma^2/(1 + 4\Gamma)^4$, involving the same contact transparency $\Gamma$ for all contacts (Di Lorenzo and Nazarov [2005]). The denominator properly normalises the conductance in the case of four leads attached to $S$. This result remains valid even in the case of finite temperature. It also gives the probabilities $P_{ij}$ for simultaneous detection of an electron at $Fi$ and $Fj$ by (Di Lorenzo and Nazarov [2005])

$$P_{ij} = G_{ij} / \sum_{\{k,l\}} G_{kl}, \tag{10.3}$$

where the sum over $\{k, l\}$ is over all pairs of detectors considered. We will now focus only on events that involve both Alice and Bob in order to obtain the correlator in Eq. (10.1), so that $i = 1, 2$ and $j = 3, 4$ and consequently disregard events in which both electrons go to the same FM or both go to Alice/Bob. Since the FMs polarisations are antiparallel for the two FMs belonging to Alice and Bob we associate the detection of an electron in $i = 1,\ j = 3$ with a measurement '+' and $i = 2,\ j = 4$ with a measurement '−'. If we only focus on the events involving both Alice and Bob we can immediately calculate the probabilities $P_{\mathbf{mm}',\pm\pm}$ since if in such an event an electron does

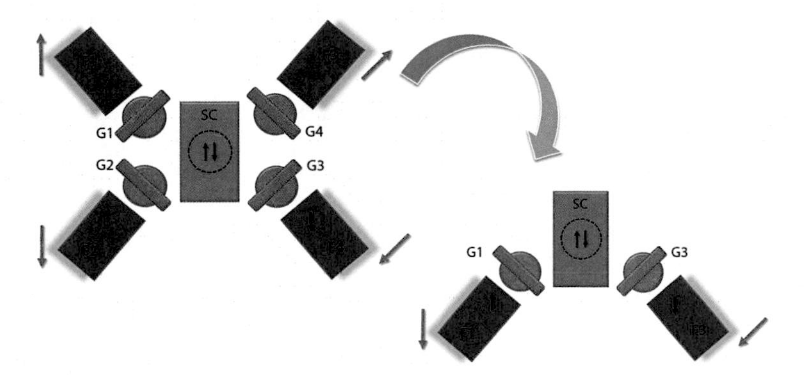


**Figure 10.2.:** Sketch of the experimental setups. **(a)**: A central SC finger $S$ (blue) is contacted to two InAs nanowires (brown). These in turn are contacted by four FM electrodes (red), the direction of magnetization of which are indicated by arrows here for the configuration of $\mathbf{m}_A$ and $\mathbf{m}_B$ as in Fig. 10.1 . Alice and Bob are represented by the spin detectors consisting of electrodes $F1$, $F2$ and $F3$, $F4$ respectively. The emerging QDs between the SC finger and the four FM electrodes are tunable by top gates $G1$-$G4$. **(b)**: Using our analysis the setup can be considerably simplified to a Y-junction geometry.

not arrive at e.g. $i = 1$ it has to go via $i = 2$ due to our choice of events. Performing the sum in Eq. (10.3) using our expression from Eq. (10.2) leads to a simple expression $E(\mathbf{m}_A, \mathbf{m}_B) = -\mathbf{g}_A\mathbf{g}_B$ for arbitrary $f(V, \delta r, \Sigma)$. Using this result the Bell parameter is given by

$$\epsilon = P^2\epsilon_0, \tag{10.4}$$

$$\epsilon_0 = |\mathbf{m}_A\mathbf{m}_B + \mathbf{m}'_A\mathbf{m}_B + \mathbf{m}'_A\mathbf{m}'_B - \mathbf{m}_A\mathbf{m}'_B|. \tag{10.5}$$

The maximum value for $\epsilon_0$ for an appropriate choice of angles (as shown in Fig. 2.3 ) between the different measurement directions is $2\sqrt{2}$. A violation of Bell's inequality is reached for $\epsilon > 2$ which requires the polarisation $P$ to be at least 84%.
The calculation above does not only apply to a tunnel contact but to all systems whose conductance has the form of Eq. (10.2). We now show that this is the case for a number of systems by considering several possible modifications of the abovementioned setup.
In Morten et al. [ 2008] diffusive charge transfer instead of ballistic transport through a tunnel contact as discussed above was considered, however, Eq. (10.2) is recovered with a modified, but constant, $f(V, \delta r, \Sigma)$. Consequently the simple form of the Bell inequality in Eq. (10.4) is recovered independent from the type of charge transfer.
In the next step we consider the resonant level setup considered in Section 6.1 in order to study effects of energy-dependent tunneling and finite voltage bias. The differential conductance at $T = 0$ for the CPS process can be directly read off the CGF in Eq. (B.2)

$$G_{R,ij}(V) = \sum_\sigma 2[T_{\mathrm{CA}ij\sigma}(V) + T_{\mathrm{CA}ij\sigma}(-V)], \tag{10.6}$$

where $T_{\mathrm{CA}ij\sigma}$ is given in Eq. (B.8). In Eq. (B.8) we assumed the polarisation of the FMs to be collinear. For the case of different directions for the polarisation we recover the same form replacing $P_i \to \mathbf{g}_i = P_i\mathbf{m}_i$ and the conductance in Eq. (10.6) is again of the form in Eq. (10.2).
Up to now we only considered a single QD not including any dependence on a finite length $\delta r$ which

was crucial for the discussion in Chapter 7 and a non-interacting system. The case of a strongly coupled triple QD was discussed in Soller and Komnik [ 2011b]. In this case one can neglect resonant tunneling and evaluate the transmission coefficients by multiplying the transmission coefficients through the different QDs. The conductance is again of the form in Eq. (10.2). Therefore, let us focus now on systems with a single dot at zero bias in order to keep expressions manageable.
First, we consider the finite length of the nanotubes/nanowires typically used in the experiments to form the QDs as done in Chapter 7 : the tunneling rate between the SC and the QD acquires a dependence on the width $\delta r$ of the SC finger of the kind $[\sin(k_F\delta r)/(k_F\delta r)]^2 \exp[-2\delta r/(\pi\xi)]$. Inlcuding these rescaled tunnel rates in Eq. (10.6) the conductance remains of the form of Eq. (10.2). We note that not coupling the SC directly to the QD but using a topological insulator which acquires a SC gap via the proximity effect of a SC slab deposited on top leads to a similar dependence on the width $l$ of the superconducting slab in the conductance expression (Adroguer et al. [ 2010]), leaving it to be of the form in Eq. (10.2).
The last example includes interactions. We follow Cuevas et al. [ 2001] and generalize the result of one normal terminal to the case of four FMs attached to a SC via a QD. In this case zero-bias conductance is given by

$$G_{I,ij} = \frac{16\tilde{\Gamma}_s^2\Gamma_F^2(1-\mathbf{g}_\alpha\mathbf{g}_\beta)}{(\tilde{\epsilon}_D^2+(4\Gamma_F)^2+\tilde{\Gamma}_s^2)^2}, \tag{10.7}$$

where $\tilde{\Gamma}_S = \Gamma_{S1} - \Sigma_{12}(0)$ and $\tilde{\epsilon}_D = \epsilon_D - \Sigma_{11}(0)$. $\Sigma_{11}(0)$ and $\Sigma_{12}(0)$ represent the 11- and 12-component of the Nambu self-energy at zero energy (Cuevas et al. [ 2001]). We do not want to specify the type of interaction any further but interactions with a local phonon as analysed in Chapter 9 would be a typical example. The conductance is again of the form of Eq. ( 10.2).
Therefore, we have shown that Eq. (10.4) is valid for a number of systems independent of the type of charge transfer, temperature, magnetic fields, spin-independent interaction or finite voltage. This important result can be explained by the fact that in Eq. (10.3) we normalise the conductances. Therefore, the overall rate of CPS processes, which is of course affected by the above mentioned effects, does not enter Eq. (10.3).
We conclude from this analysis that the non-local conductances have the form of Eq. (10.2). As they only depend on the pairwise alignment of the polarisation we can obtain all necessary non-local conductances also from just two FMs attached to the superconductor (Di Lorenzo and Nazarov [2005]). Consequently, we can also work with a typical Y-junction geometry as for the splitters indicated in Fig. 10.2 (b).
Nonetheless, polarisations much higher than $P \approx 40\%$ are hard to reach with present day materials (Coey [ 2001]). However, there is a difference between the verification of entanglement and the actual violation of a Bell inequality, which implies a violation of 'local reality' (Werner [ 1989]). Everything we need to know about the quantum state in question is whether it is not separable, meaning that it cannot be written as a convex combination of tensor product states. Generically separability implies stronger inequalities than local reality (Roy [ 2005], Uffink and Seevinck [ 2008]). For any locally realistic theory the Bell parameter in Eq. (10.1) has to be smaller than 2. Now, let us assume that Alice's and Bob's two measurement directions for spin are orthogonal, meaning e.g. that $\mathbf{m}_A$ points in $x$-direction whereas $\mathbf{m}'_A$ points in $y$-direction and the same for Bob. In this case $\epsilon_0$ in Eq. (10.5) can still reach the maximal value (Cirel'son [ 1980]) of$2\sqrt{2}$ (e.g. for $\mathbf{m}_A = \hat{\mathbf{e}}_x$, $\mathbf{m}'_A = \hat{\mathbf{e}}_y$, $\mathbf{m}_B = 1/\sqrt{2}[\hat{\mathbf{e}}_y + \hat{\mathbf{e}}_x]$ and $\mathbf{m}'_B = 1/\sqrt{2}[\hat{\mathbf{e}}_y - \hat{\mathbf{e}}_x]$, see Fig. 10.1 ). However, the maximal value for the Bell parameter in Eq. (10.1) for a separable quantum state is only $\sqrt{2}$ (Roy [ 2005]). Therefore, witnessing an entangled quantum state via Eq. (10.4) only requires $P^2 2\sqrt{2} \geq \sqrt{2} \Rightarrow P \gtrsim 70\%$. Consequently, restricting ourselves to an entanglement witness and a specific choice of measurement directions allows for a much less severe bound on the polarisation. Nonetheless, we have not made contact to a present day experiment yet.

In the last part we want to describe an actual experiment which is feasible and able to witness entanglement in a SC beamsplitter. Di Lorenzo and Nazarov [2005] were aiming at the measurement of the probabilities in (10.3) via the detection of single events in the spin detectors and averaging over them. However, these time-resolved coincidence measurements are not necessary since the probabilities can also be obtained by measuring the non-local conductances and calculating the probabilities from them. The possibility of experimental access to the non-local conductance in SC beamsplitters has been discussed in Chapter 7.
In the experiment, however, we would also need to realize a polarisation exceeding 70%. This can be realized if we remember that the independence of interactions was derived under the assumption that the interaction is not spin-specific. Indeed, if we do not operate the two QDs shown in Fig. 10.2 (b) on resonance but apply a top-gate voltage such that we are in the Kondo regime we have two SC-QD-FM devices as analysed in Chapter 8. The splitting of the Kondo resonance due to the exchange field of the FM leads to a bias tunable spin-polarisation of the current that reaches $P \approx 70\%$ as we have shown in Section 8.2.
Finally, there are new developments of nanometer sized synthetic antiferromagnets (Kubota et al. [2008], Wang et al. [2011]). These layered materials consist of coupled ferromagnetic-non-magnetic-ferromagnetic trilayered structures which have a fixed polarized ferromagnetic bottom layer but a free ferromagnetic top layer. The top layer spin polarization can be adjusted by small local magnetic fields which permits to achieve the necessary tunability of magnetisation. In principle they could also allow for fast switching, necessary to close loopholes in the Bell inequality.

## 10.2. Conclusions

We have provided three steps towards the verification of entanglement in SC beamsplitters. The first step was to calculate the Bell parameter for these devices and to show that all necessary information can be obtained from a typical Y-junction setup. In the second step we have shown that the degree of polarisation needed to verify entanglement in such experiments is much lower than the polarisation needed to violate a Bell inequality. These steps allowed us in the third step to show that such experiments can actually be realized by exploiting the latest technological advances in on-chip electronics and measurement techniques. In this way we overcome the need for measuring higher cumulants or time-resolved measurement schemes. The verification of entanglement would also mean further progress towards solid-state quantum computation.

# Chapter 11

## Conclusions and outlook

In this thesis we have considered transport properties of multi-terminal devices with special emphasis on their non-linear transport properties. The electrodes of the setups in question were of different nature: normal metals, ordinary and unconventional superconductors, ferromagnets as well as semiconductor heterostructures, systems referred to as hybrids. The terminals were supposed to be coupled together by a QPC and/or quantum dots modelled by the Anderson Hamiltonian. Due to the non-trivial interplay of different interactions (Coulomb repulsion on the dot, ferromagnetic correlations, Cooper pairing, electron-phonon coupling) the transport properties of such hybrids show a number of pronounced and very interesting effects.
We started by a detailed study of superconducting quantum point contacts to normal conductors and ferromagnets. In order to fully characterise the low-frequency electronic transport, we calculated the cumulant generating function. Especially, we considered the possibility of spin-active scattering at the superconductor-ferromagnet interface due to the interplay of the ferromagnetic exchange field in both the bulk and the interface. This scattering mechanism allows for spin-flipped Andreev reflections at superconductor-ferromagnet interfaces and gives rise to triplet correlations in the ferromagnet.
We have used these triplet correlations in order to demonstrate the possibility to generate Majorana fermions in ferromagnetic wires. The topologically non-trivial ground state is a consequence of the induced $p$-wave superconductivity and the ferromagnetic exchange field. The Majorana fermions represent zero-energy bound states in the effective $p$-wave superconductor and we found the higher cumulants of the current flow to be an ideal tool to unambiguously detect the presence of such bound states. We also studied the generation of Majorana fermions in exciton condensates.
These condensates have a finite length and so we discussed a 4-terminal setup in order to contact the two layers of the exciton condensate from two sides. A rich variety of charge transfer phenomena can be observed in such devices. As the excitons are bound states of electrons and holes in different layers they allow for perfect Coulomb drag. We found this property to be additionally reflected in a negative cross-correlation of the currents in the two layers equal in magnitude to the noise. The possibility of perfect Coulomb drag allows for nanoscale voltage transformation when coupling one layer of the exciton condensate to another mesoscopic circuit. We also investigated the possibility of a crystal phase of excitons which arises due to the dipole repulsion of the electron-hole pairs. For an intermediate range of densities we observed a regular lattice structure and also discussed its

detection via the correlation function or statistical analysis.
Concerning quantum impurity setups we first discussed non-interacting Cooper pair splitters. Apart from Cooper pair splitting we found the process of Andreev reflection enhanced transmission giving rise to a positive cross-correlation. This process transfers charge between the two normal conducting arms of the Cooper pair splitter via two Andreev reflections to the superconductor. We found such higher order processes to be relevant for highly transparent contacts. In a second step we still studied non-interacting superconductor-ferromagnet splitters but this time taking spin-active scattering into account. In this setup we observed $p$-wave Cooper pair splitting by spin-flipped crossed Andreev reflection. This process also leads to a positive cross-correlation of currents in the leads attached to the splitter as an indication of entanglement. A closer investigation of the effect of Coulomb interaction revealed that it gives rise to a variety of new transport phenomena, especially including correlated Andreev reflections which show up as quartic contributions to the cumulant generating function. Finally, we investigated the deep Kondo limit in order to make contact with contemporary experiments. This limit allows again for a non-interacting description. In contrast to previous treatments we explicitely accounted for the possibility of Andreev reflections and found their presence be directly observable in the noise. Using the same description we also discussed the possibility of a double-dot Cooper pair splitter. As for the splitters discussed before a positive cross-correlation has been found as a consequence of Cooper pair splitting.
We additionally proposed a generic model for Cooper pair splitters taking into account the specific properties of the materials for the double dot. We compared its results to experimental data and made contact to different theoretical models. We showed that the non-local conductance represents an ideal tool to investigate Cooper pair splitting. However, for highly efficient splitters we illustrated that the discrimination of Cooper pair splitting from local transport processes in the conductance becomes difficult since neither of them can be treated as a perturbation to the other.
We also investigated ferromagnet-quantum dot-superconductor junctions taking into account the properties of the materials and interaction both for even and odd charge states of the quantum dot. In the Kondo regime we showed that such hybrid devices can be used for spin-filtering since due to the Kondo effect a collective state of the dot and the leads is formed inhibiting specific interface effects. On the contrary, in an even charge state we explained a new subgap feature associated to spin-active scattering and also investigated its behavior in an applied magnetic field.
The transient dynamics and the effect of electron-phonon interaction was discussed for a resonant level model which might be relevant for Kondo dots. We used two analytical models and compared them to diagrammatic Monte Carlo simulations and obtained good agreement. We found that much stronger phonon sidebands of the Kondo resonance can be expected compared to the normal-conducting case due to the interplay of the phonon density of states on the quantum dot with the superconducting density of states in the lead.
Finally, we used our results in order to propose an experiment to witness entanglement in superconducting beamsplitters only using non-local conductance measurements.
Nonetheless, many questions remain for future works. The experimental realisation of Majorana fermions paves the way towards more involved hybrid setups, most of which are still barely understood. Likewise, the effect of electron-phonon interaction in superconductor hybrids should be investigated in full detail both by analytical means and Monte Carlo approaches. Concerning Cooper pair splitting we have seen significant progress in the last years both in theory and in experiment. Recent technological advances as structured bottom-gates will lead to a substantial improvement of our understanding of the proximity effect in nanotubes/nanowires. The realisation of exciton condensate junctions provides us with a new BCS condensate to be studied in hybrid configurations. It is tempting to speculate that replacing superconductors by exciton condensates could allow for the realisation of room-temperature quantum nanoelectronics.

# Peer reviewed Publications

Large parts of this thesis have already been published or are currently under peer-review. Papers in preparation are also listed. (Information as of March 7, 2013)

- H. Soller and A. Komnik. Charge transfer statistics and entanglement in normal-quantum dot-superconductor hybrid structures. *Eur. Phys. J. D*, 63(1):3, 2011b. doi:10.1140/epjd/e2010-00256-7

- H. Soller and A. Komnik. Hamiltonian approach to the charge transfer statistics of Kondo quantum dots contacted by a normal metal and a superconductor. *Physica E*, 44(2):425, 2011a. doi:10.1016/j.physe.2011.09.014

- H. Soller, F. Dolcini, and A. Komnik. Nanotransformation and current fluctuations in exciton condensate junctions. *Phys. Rev. Lett.*, 108:156401, 2012c. doi:10.1103/PhysRevLett.108.156401

- H. Soller, L. Hofstetter, S. Csonka, A. Levy Yeyati, C. Schönenberger, and A. Komnik. Kondo effect and spin-active scattering in ferromagnet-superconductor junctions. *Phys. Rev. B*, 85: 174512, 2012d. doi:10.1103/PhysRevB.85.174512

- H. Soller and A. Komnik. P-wave Cooper pair splitting. *Beilstein J. Nanotechnol.*, 3:493, 2012. doi:10.3762/bjnano.3.56

- H. Soller and A. Komnik. Current noise and higher order fluctuations in semiconducting bilayer systems. *Fluctuations and Noise Letters*, 12(2), 2013

# Publications in preparation

- H. Soller, J. P. Dahlhaus, and A. Komnik. Charge transfer statistics of transport through Majorana bound states. *in preparation*, 2012b

- D. Breyel, H. Soller, T. L. Schmidt, and A. Komnik. Detecting an exciton crystal by statistical means. *submitted for peer review*, 2013

- H. Soller and D. Breyel. Detecting phase transitions in excitonic systems via conductance. *in preparation*, 2013

- H. Soller, P. Burset, L. Hofstetter, A. Baumgartner, B. Braunecker, K. Kang, C. Schönenberger, A. Komnik, and A. Levy Yeyati. The generic model of Cooper pair splitting. *in preparation*, 2012a

- K. F. Albrecht, H. Soller, L. Mühlbacher, and A. Komnik. Transient dynamics and steady state behavior of the Anderson-Holstein model with a superconducting lead. *in preparation*, 2012a

- H. Soller, L. Hofstetter, and D. Reeb. Entanglement witnessing in superconducting beamsplitters. *submitted for peer review*, 2012e


## Publications not related to this thesis


- H. Soller and A. Wedemeier. Prediction of synergistic multi-compound mixtures - a generalized Colby approach. *Crop Protection*, 42:180, 2012

- H. Soller, F. Kulmann, and W. Rödder. Efficient and exact solutions for job shop scheduling problems via a generalized colouring algorithm for tree-like graphs. *submitted for peer review*, 2012f

- Henning Soller. Hg-lg mode conversion with stressed 3-mode fibers under polarization. *Open J. Appl. Sci.*, 2(4):224, 2012. doi:10.4236/ojapps.2012.24033

- H. Soller. Fcs of superconducting tunnel junctions in non-equilibrium. *Internat. J. Mod. Phys. B*, 27, 2013. URL `http://arxiv.org/abs/1302.1106`

# Appendix A

## Non-equilibrium formalism for quantum impurity systems

### A.1. Non-equilibrium Green's functions: the Keldysh formalism

This appendix serves as a short introduction to the non-equilibrium GF formalism that is used for most calculations in this thesis. Non-equilibrium quantum field theory, however, is a subject of its own and more details on the formalism for the calculation of the CGF can be found in Maier [2011] and Schmidt [2007], on the formalism for SC systems in Atienza [2012], Cuevas [1999], Hofstetter [2011] and Soller [2009] and, of course, the references therein.
In quantum mechanics the GF is often interpreted as the inverse of a differential operator. Since time evolution in quantum mechanics is always described by a differential operator we can (in principle) define GFs for any Hamiltonian. However, their calculation may be complicated, especially if the Hamiltonian is not quadratic, see Chapter 6. Therefore, in many cases it is convenient to decompose the Hamiltonian into $H = H_0 + V_{\text{per}}$ with a perturbation $V_{\text{per}}$ and a diagonal part $H_0$. $V_{\text{per}}$ may be diagonal as well (see Chapter 2) or not (see Chapter 6). In the latter case one typically assumes the perturbation to be small in some parameter.
We have seen in Chapter 2 that, in order to describe non-equilibrium situations, we need to work on the Keldysh contour introduced in Fig. 2.2. Additionally, a useful starting point for a calculation can only be an expectation value of two fermionic operators since expectation values of single operators vanish. Therefore we define GFs as two-point correlation functions of two operators $O_{1H}$ and $O_{2H}$ on the Keldysh contour as

$$G_{O_1O_2}(x,x',t,t') = -i\langle T_C O_{1H}(x,t) O_{2H}(x',t') S_C\rangle_0, \tag{A.1}$$

where $O_{1H}$, $O_{2H}$ are operators that create/annihilate particles in eigenstates of the unperturbed Hamiltonian $H_0$ and consequently the average is also taken with respect to eigenstates or mixed states of $H_0$. The interaction is included via the additional $S$-matrix which contains an integral along the Keldysh contour

$$S_C = T_C \exp\left[-i \int_C dt\, V_{\text{per}}(t)\right]. \tag{A.2}$$

The operator $T_\mathcal{C}$ defines the ordering of $t$ and $t'$ on the Keldysh contour which leads to the emergence of four different GFs depending on the location of $t$, $t'$ which are given by

$$\begin{aligned} G^{--}(x,x',t,t') &= -i\langle T_\mathcal{C} O_{1H}(x,t) O_{2H}(x',t') S_C\rangle_0, && \text{for } t,t' \in \mathcal{C}^-, \\ G^{-+}(x,x',t,t') &= i\langle O_{1H}(x,t) O_{2H}(x',t') S_C\rangle_0, && \text{for } t \in \mathcal{C}^-,\ t' \in \mathcal{C}^+, \\ G^{+-}(x,x',t,t') &= -i\langle O_{1H}(x,t) O_{2H}(x',t') S_C\rangle_0, && \text{for } t \in \mathcal{C}^+,\ t' \in \mathcal{C}^-, \\ G^{++}(x,x',t,t') &= -i\langle \tilde{T} O_{1H}(x,t) O_{2H}(x',t') S_C\rangle_0, && \text{for } t,t' \in \mathcal{C}^+, \end{aligned} \tag{A.3}$$

where we have used the anti-time ordering operator $\tilde{T}$ which brings operators in reverse order as $T$. It is common to combine the Keldysh GFs in a Keldysh matrix defined by

$$\mathbf{G} = \begin{pmatrix} G^{--} & G^{-+} \\ G^{+-} & G^{++} \end{pmatrix}. \tag{A.4}$$

Not all of the contour-ordered Keldysh GFs are independent. In fact, they satisfy the relation

$$G^{--} + G^{++} - G^{-+} - G^{+-} = 0, \tag{A.5}$$

which is known as the causality condition. This condition can be used in order to simplify the matrix by the orthogonal transformation

$$\mathbf{R} = \frac{1}{\sqrt{2}} \begin{pmatrix} 1 & 1 \\ -1 & 1 \end{pmatrix}. \tag{A.6}$$

The transformed matrix can be written as

$$\mathbf{G}' = \mathbf{R}^T \mathbf{G} \mathbf{R} = \begin{pmatrix} 0 & G^{--} - G^{+-} \\ G^{--} - G^{-+} & G^{--} + G^{++} \end{pmatrix} = \begin{pmatrix} 0 & G^A \\ G^R & G^K \end{pmatrix}. \tag{A.7}$$

However, we have pointed out in 2.1 that, in order to calculate the CGF, we have to work with contour-dependent counting fields. Introducing the contour dependence invalidates Eq. (A.5) and therefore, when concerned with the CGF, we work with the GFs in Eq. (A.4) whereas when we are only concerned with the current we can also use Eq. (A.7).

## A.2. Important Green's functions

As pointed out in Chapter 2 the starting point of every calculation are the GFs of the isolated components of the quantum impurity system. These shall be provided here for the systems we need meaning normal conductors, FMs, SCs and QDs. We will introduce the ones needed for ExCs by transforming the ones for the SC appropriately.

### A.2.1. Normal conductors

The simplest system is a free electron gas which we use to describe normal metals. This non-interacting description is justified since either local impurities in the leads are screened so that they do not matter or the impurity is a local contact and we have to treat it as a QD (see Section A.3). Since metals are spin-symmetric we consider the spinless case where the Hamiltonian is given by

$$H_1 = \sum_k \epsilon_k \Psi^+_{k1} \Psi_{k1}. \tag{A.8}$$

$\epsilon_k$ is the dispersion relation and $\Psi^+_{k1}$ and $\Psi_{k1}$ are the fermionic creation/annihilation operators used for the normal metal (see Section 3.1). The Heisenberg time-dependence of the $\Psi_{k1}$ operators is given by

$$\Psi_{k1}(t) = e^{-i\epsilon_k t}\Psi_{k1}(0). \tag{A.9}$$

Since the normal electrode is described as a Fermi gas, we know for the particle number

$$n_k = \langle \Psi^+_{k1}\Psi_{k1}\rangle_0 = n_F(\epsilon_k - \mu_1) = \frac{1}{1+e^{\beta(\epsilon_k-\mu_1)}}, \tag{A.10}$$

where $\beta = 1/T$ and the chemical potential is given by $\mu_1$. Therefore, the Keldysh GF for the normal lead is

$$\begin{aligned} g_{kk'1}(t,t') &= -i\langle T_C \Psi_{k1}(t)\Psi^+_{k'1}(t')\rangle_0 && \text{(A.11)} \\ &= -i\delta_{kk'}e^{-i\epsilon_k(t-t')}\begin{pmatrix} \theta(t-t')-n_k & -n_k \\ 1-n_k & \theta(t'-t)-n_k \end{pmatrix}. && \text{(A.12)} \end{aligned}$$

As only $k = k'$ contributes we only write $k$ in the following. Transforming to the energy domain yields

$$g_{k1}(\omega) = 2\pi i\delta(\omega-\epsilon_k)\begin{pmatrix} n_k - 1/2 & n_k \\ n_k - 1 & n_k - 1/2 \end{pmatrix} + P_V\frac{1}{\omega-\epsilon_k}\begin{pmatrix} 1 & 0 \\ 0 & -1 \end{pmatrix}, \tag{A.13}$$

where we encounter a principal value denoted by $P_V$. For the typical tunneling Hamiltonians we are interested in (see Eq. 2.3) tunneling which is local at some point $x = 0$. Therefore, we have to Fourier transform once more. The integral over momentum space involves the DOS $\rho_{01}$ as a prefactor which we assume to be constant (wide flat band limit). This assumption is justified since all physical processes involve states close to the Fermi edge. In addition, perfect agreement with experimental data reconfirms this assumption. We obtain

$$g_1(x=0,\omega) = 2\pi i\rho_{01}\begin{pmatrix} n_{1+} - 1/2 & n_{1+} \\ n_{1+} - 1 & n_{1+} - 1/2 \end{pmatrix}, \tag{A.14}$$

where $n_{1+}$ is a shorthand notation for $n_F(\omega - \mu_1)$. From now on we drop the additional variable $x = 0$.

### A.2.2. Ferromagnets

FMs are basically described by the same Hamiltonian as normal leads (see Eq. 3.21), however, including the exchange energy which leads to a different DOS for spin-$\uparrow$ and spin-$\downarrow$. We therefore lift the spin degeneracy that was present in Section A.2.1 and have to introduce two GFs for the different spin species

$$g_{F\sigma}(\omega) = 2\pi(1+\sigma P)i\rho_{0F}\begin{pmatrix} n_{F+} - 1/2 & n_{F+} \\ n_{F+} - 1 & n_{F+} - 1/2 \end{pmatrix}, \tag{A.15}$$

where $n_{F+}$ refers to the Fermi distribution for the FM and $P$ represents the polarisation of the FM.

### A.2.3. BCS Condensates

The Hamiltonian for the SC in Eq. (2.7) is not diagonal in the typical sense of Eq. (A.8). However, we can go over to a diagonal basis by means of Bogoliubov transformation. We define new fermionic operators

$$B^+_{k\uparrow} = u_k c^+_{k\uparrow} - v_k c_{-k\downarrow},\ B^+_{-k\downarrow} = u_k c^+_{-k\downarrow} + v_k c_{k\uparrow}, \tag{A.16}$$

and their adjoints. These operators are ordinary fermionic operators as long as the normalization condition $u_k^2 + v_k^2 = 1$ is fulfilled. This can easily be done choosing $u_k = \cos\phi$, $v_k = \sin\phi$ such that $\tan(2\phi) = -\Delta/\epsilon_k$. The operators $B_{k\sigma}$ are referred to as Bogoliubov quasiparticle annihilation operators. We immediately observe that for $\epsilon_k = 0$

$$B_{0\downarrow} = \frac{1}{\sqrt{2}}(c^+_{0\uparrow} + c_{0\downarrow}). \tag{A.17}$$

If we had equal spins as in Eq. (2.11) and not opposite ones as in the above Eq. (A.17) $B_{0\downarrow}$ would correspond to a Majorana fermion. This property is discussed in more detail in Chapter 4.
Using Eq. (A.16) on Eq. (2.7) leads to

$$H_S = \sum_{k,\sigma} E_k B^+_{k\sigma} B_{k\sigma}, \tag{A.18}$$

where $E_k = \sqrt{\epsilon_k^2 + \Delta^2}$. For this Hamiltonian we have shown how to find the corresponding GFs $g^{B_\sigma}_{kk'}$ in Section A.2.1. However, working with the transformed operators $B_{k\sigma}$ is not always the best choice, since they represent coherent superpositions of electrons and holes. Often it is advantageous to transform back to the original operators. However, we already see from Eq. (2.7) that the price to pay is the introduction of anomalous GFs for the SC electrode

$$f_{kk'S}(t,t') = i\langle T_\mathcal{C} c_{-k\downarrow}(t) c_{k'\uparrow}(t')\rangle_0,\ f^+_{kk'S}(t,t') = i\langle T_\mathcal{C} c^+_{k\uparrow}(t) c^+_{-k'\downarrow}(t')\rangle_0. \tag{A.19}$$

For simplicity we can set $t' = 0$ since the problem is time translation invariant. The GFs for the original operators can be expressed in terms of the GFs $g^{B_\sigma}_{kk'}$ introduced before via

$$\begin{aligned} g_{kk'S\sigma}(t) &= u_k^2 g^{B_\sigma}_{kk'}(t) + v_k^2 g^{B_{-\sigma}}_{kk'}(-t), &\text{(A.20)}\\ f_{kk'S}(t) &= f^+_{kk'S}(t) = -u_k v_k [g^{B_{-\sigma}}_{kk'}(t) + g^{B_\sigma}_{kk'}(-t)]. &\text{(A.21)} \end{aligned}$$

Formally we can combine the three GFs and the time-reversed GF for the holes into one matrix of GFs which corresponds to the GF in Nambu space

$$\mathbf{G}_S(\omega) = \begin{pmatrix} g_S(\omega) & f_S(\omega) \\ f^+_S(\omega) & g_S(-\omega) \end{pmatrix}. \tag{A.22}$$

We can calculate this GF using the same steps as in Section A.2.1 and arrive at (Jonckheere et al. [2009])

$$\begin{aligned} \mathbf{G}_S(\omega) &= \pi\rho_{0S}\{\omega\tilde{\mathbf{1}} - \Delta\tilde{\sigma}_x\}\left\{ i\,\mathrm{sign}(\omega)\frac{\theta[(|\omega|-\Delta)/\Delta]}{\sqrt{\omega^2-\Delta^2}} \begin{pmatrix} 2n_S - 1 & -2n_S \\ 2n_S - 2 & 2n_S - 1 \end{pmatrix} \right. \\ &\quad \left. -\frac{\theta[(\Delta-|\omega|)/\Delta]}{\sqrt{\Delta^2-\omega^2}} \begin{pmatrix} 1 & 0 \\ 0 & -1 \end{pmatrix} \right\}, \end{aligned} \tag{A.23}$$

where the matrices $\tilde{1}$ and $\tilde{\sigma}_x$ act in Nambu space. Taking the limit of small energy we arrive at Eq. (3.14) whereas in the high energy limit we obtain Eq. (3.15).
In the case of an ExC the Hamiltonian in Eq. (2.10) is very similar to the BCS Hamiltonian for SCs. Consequently, the GFs are defined analogous to Eq. (A.11) and (A.19) with the transformation of $\uparrow$-electrons to $T$-electrons and $\downarrow$-electrons to $B$-holes. For low energy $\omega \ll \Delta$ only the anomalous GF in the ExC will contribute and we arrive at

$$f_{\text{ExC}}(\omega) = \pi \frac{\Delta \rho_{0E}}{\sqrt{\Delta^2 - \omega^2}} \begin{pmatrix} 1 & 0 \\ 0 & -1 \end{pmatrix}, \; g_{\text{ExC,T,B}}(\omega) = 0. \tag{A.24}$$

In the limit of high energy $\omega \gg \Delta$ only the normal GF in the ExC will contribute. Additionally, we have to mind that the top and bottom layer of the ExC may have different chemical potentials so that

$$f_{\text{ExC}}(\omega) = 0, \; g_{\text{ExC,T,B}}(\omega) = i\pi \frac{\omega \rho_{0E}}{\sqrt{\omega^2 - \Delta^2}} \begin{pmatrix} 2f_{T,B} - 1 & 2f_{T,B} \\ 2f_{T,B} - 2 & 2f_{T,B} - 1 \end{pmatrix}, \tag{A.25}$$

where $f_{T,B}$ refer to the Fermi distributions for the top- and bottom-layer respectively.

## A.3. Fermionic Dot Level

The final simple GF we want to introduce is that of a fermionic level. In this case we have the Hamiltonian

$$H_D = \sum_\sigma \epsilon_{D\sigma} d_\sigma^+ d_\sigma. \tag{A.26}$$

The Heisenberg equation of motion leads to $d_\sigma(t) = e^{-i\epsilon_{D\sigma}t} d_\sigma(0)$ and the time-dependent GF can easily be shown to be

$$D_\sigma(t) = -i\langle T_C d_\sigma(t) d_\sigma^+(0)\rangle_0 = -ie^{-i\epsilon_{D\sigma}t} \begin{pmatrix} \theta(t) - n_\sigma & -n_\sigma \\ 1 - n_\sigma & \theta(-t) - n_\sigma \end{pmatrix}, \tag{A.27}$$

where $n_\sigma = \langle d_\sigma^+ d_\sigma \rangle_0$ is the ground state occupation number of the dot for spin $\sigma$. We assume that the steady state of the system does not depend on the initial occupation $n_\sigma$ which is fulfilled in all cases we want to consider (Albrecht et al. [2012b]) and we find

$$D_\sigma(\omega) = \begin{pmatrix} 1/(\omega - \epsilon_{D\sigma}) & 0 \\ 0 & -1/(\omega - \epsilon_{D\sigma}) \end{pmatrix}. \tag{A.28}$$

This is also one of the GFs that are used throughout this work.

# Appendix B

## Full cumulant generating functions

In this Appendix we will give several results for the full CGF that were left out of the main text. For numerical evaluation we always use the full result.

### B.1. Superconductor-ferromagnet quantum point contact

The result in Eq. (3.22) is an approximate result for the CGF of the SC-FM QPC. Here we give the full result

$$\begin{aligned}
&\ln \chi_{\mathrm{SF,full}}(\lambda) = \\
&\tau \int \frac{d\omega}{2\pi} \Bigg[ \sum_{\sigma} \ln \Bigg( \prod_{\alpha=\pm} [1 + \tilde{T}_{eF\sigma} A_\alpha(\omega,\lambda)] + \tilde{T}_{AF2}(2n_S - 1)\{(2n_S - 1)[(e^{i\lambda} - 1)^2 n_{F-}(1 - n_{F+}) \\
&-2(e^{i\lambda} - 1)(e^{-i\lambda} - 1) n_{F-} n_{F+} + (e^{-i\lambda} - 1)^2 n_{F+}(1 - n_{F-})] + 2n_S(e^{i\lambda} - 1)(e^{-i\lambda} - 1)(n_{F+} - 1 + n_{F-})\} \\
&+\tilde{T}_{BCF}\left\{(2n_S - 1)^2(e^{i\lambda} - e^{-i\lambda})^2[n_{F-}e^{i\lambda} + n_{F+}e^{-i\lambda} + \Gamma_e(1 - \sigma P) n_S(1 - n_S)(e^{i\lambda} - e^{-i\lambda})^2]\right. \\
&+n_S(2n_S - 1)\left\{4(n_S - 1)(n_{F+} - 1 + n_{F-})(e^{i\lambda} - 1 - e^{-i\lambda})^2 + \sigma P\{8[(e^{i\lambda} - 1)^2 n_{F+} - (e^{-i\lambda} - 1)^2 n_{F-}]\right. \\
&-(e^{-i\lambda} - 1)^3[e^{3i\lambda}(2n_S - 1)(1 + n_{F-} - n_{F+}) - (2n_S - 1)(n_{F-} - n_{F+} - 1) + e^{2i\lambda}(2n_S(3 + n_{F+} - n_{F-}) \\
&\quad -3 + 7n_{F+}) - e^{i\lambda}(3 + n_{F+} + 2n_S(n_{F+} - 3 - n_{F-}) + 7n_{F-})]\} + 2\sigma P[\sum_{\alpha=\pm} \alpha A_\alpha(\omega,\lambda)] \Bigg\} \theta\left(\frac{|\omega| - \Delta}{\Delta}\right) \Bigg) \\
&+ \ln\{1 + T_{AF}[n_{F+}(1 - n_{F-})(e^{2i\lambda} - 1) + n_{F-}(1 - n_{F+})(e^{-2i\lambda} - 1)]\}\theta\left(\frac{\Delta - |\omega|}{\Delta}\right) \Bigg],
\end{aligned} \tag{B.1}$$

involving the abbreviation $A_\alpha(\omega,\lambda) = \left[n_{F\alpha}(1 - n_S)(e^{i\alpha\lambda} - 1) + n_S(1 - n_{F\alpha})(e^{-i\alpha\lambda} - 1)\right]$ and the effective transmission coefficients

$$\begin{aligned}
T_{\mathrm{eF}\sigma}(\omega) &= \frac{4\Gamma_e(1 + \sigma P)}{[1 + \Gamma_A(1 + \sigma P)]^2 - \Gamma_A^2(1 - P)(1 + P)}, \\
T_{\mathrm{AF2}}(\omega) &= \frac{4\Gamma_A^2(1 + P)(1 - P)}{\{[1 + \Gamma_e(1 + P)]^2 - \Gamma_2^2(1 - P)(1 + P)\}\{[1 + \Gamma_e(1 - P)]^2 - \Gamma_A^2(1 - P)(1 + P)\}} = \frac{T_{\mathrm{BCF}}}{\Gamma_e},
\end{aligned}$$

and

$$T_{\mathrm{AF}}(\omega) = \frac{4\Gamma_A^2(1+P)(1-P)}{\Gamma_A^4(1-P^2)^2+\Gamma_A^2(1-P^2)[2-\Gamma_A^2(1+P)^2-\Gamma_e^2(1-P)^2]+[1+\Gamma_e^2(1+P)^2][1+\Gamma_e^2(1-P)^2]}.$$

This is the result we use for numerical evaluations.

## B.2. Normal-quantum dot-supercondcutor junction

The scattering matrix of a normal metal-QD-SC junction has two energy regimes (Schwab and Raimondi [1999]) which is also reflected in two energy regimes for the CGF as in Eq. (3.11). The respective Fermi distributions of the individual terminals are abbreviated by $n_{\alpha+}$ and $n_{\alpha-} := 1 - n_{\alpha+}(-\omega)$ for hole-like contributions. Since we have the superconductor in equilibrium $n_S = n_{S+} = n_{S-}$ and the result reads

$$\begin{aligned}
&\ln\chi_{\mathrm{SD}}(\boldsymbol{\lambda}) = \frac{\tau}{\pi}\int d\omega \left[\theta\left(\frac{|\omega|-\Delta}{\Delta}\right)\right.\\
&\times\left(\sum_\sigma \ln\left\{1+\sum_{i,j=1,\dots,N,S,\,i\neq j} T_{ij\sigma}(\omega)n_{i+}(1-n_{j+})(e^{i(\lambda_i-\lambda_j)}-1)\right\}\right)\\
&+\frac{1}{2}\theta\left(\frac{\Delta-|\omega|}{\Delta}\right)\left(\sum_\sigma \ln\left\{\left[1+\sum_{i,j=1,\dots,N,\,i\neq j} T_{Aij\sigma e}n_{i+}(1-n_{j+})(e^{i(\lambda_i-\lambda_j)}-1)\right]\right.\right.\\
&\times\left[1+\sum_{i,j=1,\dots,N,\,i\neq j} T_{Aij\sigma h}n_{j+}(1-n_{i+})(e^{i(\lambda_i-\lambda_j)}-1)\right]\\
&+\sum_{i=1,\cdots,N} T_{Ai\sigma}\left[n_{i+}(1-n_{i-})(e^{2i(\lambda_i-\lambda_S)}-1)\right]+\sum_{i,j=1,\dots,N,\,i\neq j} T_{\mathrm{CA}ij\sigma}\\
&\left.\left.\left.\times\left[n_{j-}(1-n_{i+})(e^{i(2\lambda_S-\lambda_i-\lambda_j)}-1)+n_{i+}(1-n_{j-})(e^{-i(2\lambda_S-\lambda_i-\lambda_j)}-1)\right]\right\}\right)\right], \qquad \text{(B.2)}
\end{aligned}$$

where we define

$$T_{ij\sigma}(\omega) = 4\Gamma_i\Gamma_j(1+\sigma P_i)(1+\sigma P_j)/\left\{(\omega-\delta_\sigma)^2+\left[\sum_{k=1,\dots,N}\Gamma_k(1+\sigma P_k)+\Gamma_{S1}\right]^2\right\}, \qquad \text{(B.3)}$$

$$T_{Aij\sigma e} = 4\Gamma_i(1+\sigma P_i)\Gamma_j(1+\sigma P_j)\left\{(\omega-\delta_{-\sigma})^2+\left[\sum_{k=1,\dots,N}\Gamma_k(1-\sigma P_k)\right]^2\right\}/W_2, \qquad \text{(B.4)}$$

$$T_{Aij\sigma h} = 4\Gamma_i(1-\sigma P_i)\Gamma_j(1-\sigma P_j)\left\{(\omega-\delta_\sigma)^2+\left[\sum_{k=1,\dots,N}\Gamma_k(1+\sigma P_k)\right]^2\right\}/W_2, \qquad \text{(B.5)}$$

$$T_{Ai\sigma} = 4\Gamma_i^2(1-\sigma P_i)(1+\sigma P_i)\Gamma_{S2}^2/W_2(\omega), \qquad \text{(B.6)}$$

(B.7)

$$
\begin{aligned}
T_{\mathrm{CA}ij\sigma} &= 4\Gamma_i(1+\sigma P_i)\Gamma_j(1-\sigma P_j)\Gamma_{S2}^2/W_2, &\text{(B.8)}\\
W_2 &= (\omega-\epsilon_{D\sigma})^2(\omega-\epsilon_{D-\sigma})^2+\left\{\left[\sum_{k=1,\dots,N}\Gamma_k(1+\sigma P_k)\right]^2+\Gamma_{S2}^2\right\}\\
&\times\left\{\left[\sum_{k=1,\dots,N}\Gamma_k(1-\sigma P_k)\right]^2+\Gamma_{S2}^2\right\}+\left[\sum_{k=1,\dots,N}\Gamma_k(1+\sigma P_k)\right]^2(\omega-\epsilon_{D-\sigma})^2\\
&+\left[\sum_{k=1,\dots,N}(1-\sigma P_k)\right]^2(\omega-\epsilon_{D\sigma})^2+2\Gamma_{S2}^2(\omega-\epsilon_{D\sigma})(\omega-\epsilon_{D-\sigma}), &\text{(B.9)}
\end{aligned}
$$

with $P_S=0$. The abbreviation $\Gamma_i=\pi\rho_{0i}|\gamma_i|^2/2$ is the (energy-independent) dot-lead contact transparency with dimension energy for the normal/FM leads. For the SC lead it is affected by the energy-dependent superconducting DOS so that

$$
\Gamma_S = \pi\rho_{0S}|\gamma_S|^2|\omega|/(2\sqrt{\omega^2-\Delta^2}),\ \Gamma_{S2}=\pi\rho_{0S}|\gamma_S|^2|\Delta|/(2\sqrt{\Delta^2-\omega^2}). \qquad \text{(B.10)}
$$

This quantity is discussed in greater detail in Section 6.1.
In Section 6.3 we discuss the interacting case. For the calculation we need the GFs in the two-terminal case with $\lambda=\lambda_1-\lambda_S$ using the approximated GFs in Eqs. (3.14) and (3.15). The GFs for $|\omega|>\Delta$ are given by

$$
\begin{aligned}
\mathcal{D}_{0\sigma}^{\lambda}(\omega) &= \frac{1}{\det A_{D1}}\left(\begin{array}{c}(\omega-\epsilon_{D\sigma})+2i\Gamma_1(n_1-1/2)+2i\Gamma_S(n_S-1/2)\\ 2i\Gamma_1e^{-i\lambda/2}(n_{1+}-1)+2i\Gamma_Se^{i\lambda/2}(n_S-1)\end{array}\right.\\
&\qquad\left.\begin{array}{c}2i\Gamma_Se^{-i\lambda/2}n_S+2i\Gamma_1e^{i\lambda/2}n_{1+}\\ -(\omega-\epsilon_{D\sigma})+2i\Gamma_S(n_S-1/2)+2i\Gamma_1(n_1-1/2)\end{array}\right),\\
\mathcal{C}^{\lambda}(\omega) &= \mathcal{C}^{\lambda+}(\omega)=0, &\text{(B.11)}
\end{aligned}
$$

with the determinant

$$
\begin{aligned}
\det A_{D1} &= (\omega-\epsilon_{D\sigma})^2+(\Gamma_1+\Gamma_S)^2+4\Gamma_1\Gamma_Sn_{1+}(1-n_S)(e^{i\lambda}-1)\\
&+4\Gamma_1\Gamma_Sn_S(1-n_{1+})(e^{-i\lambda}-1).
\end{aligned}
$$

For $|\omega|<\Delta$ we have

$$
\begin{aligned}
\mathcal{C}_0^{\lambda}(\omega) = \frac{1}{\det A_{D2}}&\left(\begin{array}{c}2\Gamma_1\Gamma_{S2}\{e^{i\lambda}n_{1+}(\epsilon_{D\uparrow}+i(1-2n_{1-})\Gamma_1-\omega)\\ +e^{-i\lambda}[n_{1-}(-\epsilon_{D\downarrow}+i(2n_{1+}-1)\Gamma_1+\omega)]\}\\ i\Gamma_{S2}(4e^{-3/2i\lambda}(n_{1+}-1)n_{1-}\Gamma_1^2-e^{i\lambda/2}\{\Gamma_{S2}^2+(\epsilon_{D\downarrow}+i(2n_{1+}-1)\Gamma_1-\omega)\\ \times[\epsilon_{D\uparrow}+i(1-2n_{1-})\Gamma_1-\omega]\})\end{array}\right.\\
&\left.\begin{array}{c}i\Gamma_{S2}(4e^{i3/2\lambda}n_{1+}(1-n_{1-})\Gamma_1^2+e^{-i\lambda/2}\{\Gamma_{S2}^2+[\epsilon_{D\downarrow}+i(1-2n_{1+})\Gamma_1-\omega]\\ \times[\epsilon_{D\uparrow}+i(2n_{1-}-1)\Gamma_1-\omega]\})\\ 2\Gamma_1\Gamma_{S2}\{e^{i\lambda}(1-n_{1-})[\epsilon_{D\downarrow}+i(2n_{1+}-1)\Gamma_1-\omega]\\ +e^{-i\lambda}(n_{1+}-1)[\epsilon_{D\uparrow}+i(2n_{1-}-1)\Gamma_1-\omega]\}\end{array}\right),
\end{aligned}
$$

$$
\mathcal{D}^{\lambda}_{0\sigma}(\omega) = \frac{1}{\det A_{D2}}
\times \begin{pmatrix}
i\Gamma_1(2n_{1+}-1)(\epsilon^2_{D-\sigma}+\Gamma_1^2) - (\epsilon_{D-\sigma}+i(1-2n_{1-})\Gamma_1)\Gamma^2_{S2} - \epsilon_{D\sigma}(\Gamma_1^2+(\epsilon_{D-\sigma}-\omega)^2) \\
+(\epsilon^2_{D-\sigma}+2i\epsilon_{D-\sigma}(2n_{1+}-1)\Gamma_1+\Gamma_1^2+\Gamma^2_{S2})\omega + (i(2n_{1+}-1)\Gamma_1 - 2\epsilon_{D-\sigma})\omega^2+\omega^3 \\
2i\Gamma_1(e^{i3/2\lambda}(n_{1-}-1)\Gamma^2_{S2} + e^{-i3/2\lambda}(n_{1+}-1)(\Gamma_1^2+(\epsilon_{D-\sigma}-\omega)^2)) \\
2i\Gamma_1(e^{-i3/2\lambda}n_{1-}\Gamma^2_{S2} + e^{i\lambda/2}(\Gamma_1^2+(\epsilon_{D-\sigma}-\omega)^2)) \\
i\Gamma_1((2n_{1+}-1)\Gamma_1^2+(2n_{1-}-1)\Gamma^2_{S2}) + \epsilon_{D\sigma}(\Gamma_1^2+(\epsilon_{D-\sigma}-\omega)^2) + i\epsilon^2_{D-\sigma}(-\Gamma_1+2n_{1+}\Gamma_1+i\omega) \\
-(\Gamma_1^2+\Gamma^2_{S2})\omega + i(2n_{1-}-1)\Gamma_1\omega^2 - \omega^3 + \epsilon_{D-\sigma}(\Gamma^2_{S2}+2\omega(i(1-2n_{1+})\Gamma_1+\omega))
\end{pmatrix},
$$

$$
\mathcal{C}^{\lambda+}_0(\omega) = \frac{1}{\det A_{D2}}
\begin{pmatrix}
2\Gamma_1\Gamma_{S2}\{e^{i\lambda}(n_{1-}-1)[\epsilon_{D\uparrow}+i(1-2n_{1+})\Gamma_1-\omega] \\
+e^{-i\lambda}(1-n_{1+})[\epsilon_{D\downarrow}+i(1-2n_{1-})\Gamma_1-\omega]\} \\
i\Gamma_{S2}(4e^{-i3/2\lambda}(n_{1+}-1)n_{1-}\Gamma_1^2 - e^{i\lambda/2}\{\Gamma^2_{S2}+[\epsilon_{D\uparrow}+i(1-2n_{1+})\Gamma_1-\omega] \\
\times[\epsilon_{D\downarrow}+i(2n_{1-}-1)\Gamma_1-\omega]\}) \\
i\Gamma_{S2}(4e^{i3/2\lambda}n_{1+}(1-n_{1-})\Gamma_1^2 + e^{-i\lambda/2}\{\Gamma^2_{S2}+[\epsilon_{D\uparrow}+i(2n_{1+}-1)\Gamma_1-\omega] \\
\times[\epsilon_{D\downarrow}+i(1-2n_{1-})\Gamma_1-\omega]\}) \\
2\Gamma_1\Gamma_{S2}\{e^{-i\lambda}n_{1-}[\epsilon_{D\uparrow}+i(2n_{1+}-1)\Gamma_1-\omega] \\
+e^{i\lambda}n_{1+}[-\epsilon_{D\downarrow}+i(1-2n_{1-})\Gamma_1+\omega]\}
\end{pmatrix}, \tag{B.12}
$$

with the determinant

$$
\begin{aligned}
\det A_{D2} &= (\omega-\epsilon_{D\uparrow})^2(\omega-\epsilon_{D\downarrow})^2 + (\Gamma_1^2+\Gamma^2_{S2})^2 + \Gamma_1^2[(\omega-\epsilon_{D\downarrow})^2+(\omega-\epsilon_{D\uparrow})^2] \\
&\quad +2\Gamma^2_{S2}(\omega-\epsilon_{D\uparrow})(\omega-\epsilon_{D\downarrow}) + 4(e^{2i\lambda}-1)n_{1+}(1-n_{1-})\Gamma_1^2\Gamma^2_{S2} \\
&\quad +4(e^{-2i\lambda}-1)n_{1-}(1-n_{1+})\Gamma_1^2\Gamma^2_{S2}.
\end{aligned}
$$

These are used as the starting point for perturbation theory.

## B.3. CGF for the FM-QD-SC junction with spin-active scattering

We use the approximation for the GFs introduced in Eqs. (3.14) and (3.15) to simplify the expression. In this approximation the CGF for a FM-QD-SC junction may be expressed as

$$
\begin{aligned}
&\ln\chi_{\mathrm{RSFa}}(\lambda) = \\
&2\tau\int\frac{d\omega}{2\pi}\Bigg[\Bigg(\ln\Bigg\{1+\Bigg[\sum_\sigma T_{\mathrm{Re}\sigma}\Bigg][n_{F+}(1-n_S)(e^{i\lambda}-1)+n_S(1-n_{F+})(e^{-i\lambda}-1)] \\
&+T_{\mathrm{Rd}}[n_{F+}(1-n_S)(e^{i\lambda}-1)+n_S(1-n_{F+})(e^{-i\lambda}-1)]^2 \\
&-T_{\mathrm{Rs}}[n_{F+}(1-n_S)(e^{i\lambda}-1)+n_S(1-n_{F+})(e^{-i\lambda}-1)]\Bigg\}\Bigg)\,\theta\left(\frac{|\omega|-\Delta}{\Delta}\right) \\
&+\frac{1}{2}\Big(\ln\{1+2T_{\mathrm{RA}}[(e^{2i\lambda}-1)n_{F+}(1-n_{F-})+(e^{-2i\lambda}-1)n_{F-}(1-n_{F+})] \\
&+T_{\mathrm{RAd}}[(e^{2i\lambda}-1)n_{F+}(1-n_{F-})+(e^{-2i\lambda}-1)n_{F-}(1-n_{F+})]^2+(T_{\mathrm{RAT}}+T_{\mathrm{RA2}}) \\
&\times[(e^{2i\lambda}-1)n_{F+}(1-n_{F-})+(e^{-2i\lambda}-1)n_{F-}(1-n_{F+})]\}\Big)\,\theta\left(\frac{\Delta-|\omega|}{\Delta}\right)\Bigg],
\end{aligned} \tag{B.13}
$$

where we set $\lambda_1 - \lambda_S =: \lambda$ and we have the transmission coefficients

$$\begin{aligned}
T_{\mathrm{Re}\sigma} &= \frac{4\Gamma_{F\sigma}(\Gamma_{S11}+\Gamma_{S12})[(\Gamma_{F-\sigma}+\Gamma_{S11}+\Gamma_{S12})^2+(\omega-\delta)^2-\Gamma_{S13}^2]}{\det A_{R10}},\\
T_{\mathrm{Rd}} &= \frac{16\Gamma_{F\uparrow}\Gamma_{F\downarrow}[(\Gamma_{S11}+\Gamma_{S12})^2-\Gamma_{S13}^2]}{\det A_{\mathrm{R10}}},\ T_{\mathrm{Rs}} = \frac{16\Gamma_{F\uparrow}\Gamma_{F\downarrow}}{\det A_{\mathrm{R10}}},\\
\det A_{\mathrm{R10}} &= [(\Gamma_{F\uparrow}+\Gamma_{S11}+\Gamma_{S12})^2+\omega^2-\Gamma_{S13}^2][(\Gamma_{F\downarrow}+\Gamma_{S11}+\Gamma_{S12})^2+\omega^2-\Gamma_{S13}^2]\\
&\quad+\Gamma_{S13}^2[(\Gamma_{F\uparrow}-\Gamma_{F\downarrow})^2+4\omega^2],\ T_{\mathrm{RA}} = \frac{4\Gamma_{F\uparrow}\Gamma_{F\downarrow}(\Gamma_{S21}+\Gamma_{S22})^2}{\det A_{\mathrm{R20}}},\\
T_{\mathrm{RAd}} &= \frac{16\Gamma_{F\uparrow}^2\Gamma_{F\downarrow}^2((\Gamma_{S21}+\Gamma_{S22})^2-\Gamma_{S23}^2)^2}{(\det A_{\mathrm{R20}})^2},\\
T_{\mathrm{RAT}} &= \frac{4\Gamma_{S23}^2[(\Gamma_{F\uparrow}^2+\Gamma_{F\downarrow}^2)(\Gamma_{S23}^2-(\Gamma_{S21}+\Gamma_{S22})^2+\omega^2)^2]}{(\det A_{\mathrm{R20}})^2}\\
T_{\mathrm{RA2}} &= \left\{4\Gamma_{S23}^2\Gamma_{F\uparrow}^2\Gamma_{F\downarrow}^2(\Gamma_{F\uparrow}^2+\Gamma_{F\downarrow}^2+4(\Gamma_{S23}^2+\omega^2))-2\Gamma_{F\uparrow}\Gamma_{F\downarrow}(\Gamma_{S21}+\Gamma_{S22})^2(\Gamma_{F\uparrow}^2+\Gamma_{F\downarrow}^2\right.\\
&\quad\left.+4(\Gamma_{S23}^2+\omega^2))\right\}/(\det A_{\mathrm{R20}})^2,\\
\det A_{\mathrm{R20}} &= \omega^4+[(\Gamma_{S21}+\Gamma_{S22})^2+\Gamma_{F\uparrow}\Gamma_{F\downarrow}-\Gamma_{S23}^2]^2+2(\Gamma_{S21}+\Gamma_{S22})^2\omega^2\\
&\quad+\Gamma_{F\uparrow}^2\omega^2+\Gamma_{F\downarrow}\omega^2+\Gamma_{S23}^2[(\Gamma_{F\uparrow}+\Gamma_{F\downarrow})^2+2\omega^2].
\end{aligned}$$

Again we have used several abbreviations in these definitions

$$\begin{aligned}
\Gamma_{F\sigma} &= \Gamma_1(1+\sigma P),\ \Gamma_{S11} = \frac{\Gamma_{\mathrm{RS1}}|\omega|}{\sqrt{\omega^2-\Delta^2}},\ \Gamma_{S12} = \frac{\Gamma_{\mathrm{RS2}}|\omega|}{\sqrt{\omega^2-\Delta^2}},\ \Gamma_{S13} = \frac{2(\Gamma_{\mathrm{RS1}}\Gamma_{\mathrm{RS2}})^{\frac{1}{2}}|\omega|}{\sqrt{\omega^2-\Delta^2}},\\
\Gamma_{S21} &= \frac{\Gamma_{\mathrm{RS1}}\Delta}{\sqrt{\Delta^2-\omega^2}},\ \Gamma_{S22} = \frac{\Gamma_{\mathrm{RS2}}\Delta}{\sqrt{\Delta^2-\omega^2}},\ \Gamma_{S23} = \frac{2(\Gamma_{\mathrm{RS1}}\Gamma_{\mathrm{RS2}})^{\frac{1}{2}}\Delta}{\sqrt{\Delta^2-\omega^2}}
\end{aligned}$$

with $\Gamma_{\mathrm{RS1}} = \pi\rho_{0S}\gamma_{S1}^2/2$ and $\Gamma_{\mathrm{RS2}} = \pi\rho_{0S}\gamma_{S2}^2/2$.
The result may be interpreted as the result for the SC-FM QPC in Eq. (3.26). Above the gap $T_{\mathrm{Re}\sigma}$ describes single electron transfer without spin flip and $T_{\mathrm{Rd}}$ describes the consecutive transfer of two electrons with different spin (again without spin flip). $T_{\mathrm{Rs}}$ refers to spin flip transmission of single electrons. Below the gap $T_{\mathrm{RA}}$ and $T_{\mathrm{RA2}}$ describe single Andreev reflection and $T_{\mathrm{RAd}}$ describes two consecutive Andreev reflections intiated by electrons with opposite spin. Spin-flipped Andreev reflection is given by $T_{\mathrm{RAT}}$.

# Appendix C

## T-matrix approach to Cooper pair splitting

This appendix discusses the $T$-matrix approach to CPS. The first part discusses the calculation of the conductances from the $T$-matrix and the second part discusses the approximation used for the effects of finite voltage.

We assume the double dot to be realized by a semiconducting nanowire as e.g. InAs. Conductance at small bias $U_i$ between the different leads $i = N1$, SC, $N2$ can be calculated via (Ferry and Goodnick [1997])

$$G(U_i) = \frac{4e^2}{h} |\langle f | T(\epsilon_i) | i \rangle|^2, \text{ where} \tag{C.1}$$

$$T(\epsilon_i) = H_{\text{TDD}} \frac{1}{\epsilon_i + eU_i + i\eta - H} (\epsilon_i - H_0), \tag{C.2}$$

is the on-shell transmission or $T$-matrix, with $\eta$ being a small positive real numer that we take to zero at the end of the calculation. Since we have to describe the initial and final states, this model allows to prescribe the possible transport processes. The $T$-matrix can be written as a power series in the tunnel Hamiltonian $H_{\text{TDD}}$

$$T(\epsilon_i) = H_{\text{TDD}} + H_{\text{TDD}} \sum_{n=1}^{\infty} \left( \frac{1}{\epsilon_i + eU_i + i\eta - H_0} H_{\text{TDD}} \right)^n . \tag{C.3}$$

We approach the non-local conductance properties of the double dot Cooper pair splitter in two steps. First we consider the limit of small coupling to the SC and small bias. Later we will relax these restrictions. In (Recher et al. [2001]) this calculation has been performed for the initial and final state of CPS in the regime

$$\Delta,\, U_{1,2} > eU_{N1,\,N2,\,\text{SC}}, \Gamma_l,\; k_B T > \Gamma_{\text{DS}l}. \tag{C.4}$$

The result for the conductance has the form

$$G'_{\text{CPS}} = \frac{4e^2}{h} \frac{4\Gamma_{\text{DS1}}\Gamma_{\text{DS2}}\Gamma_1\Gamma_2}{[(\epsilon_{D1} + \epsilon_{D2})^2 + (\Gamma_1 + \Gamma_2)^2]^2} \left[ \frac{\sin(k_F \delta r)}{k_F \delta r} \right]^2 \exp\left( -\frac{2\delta r}{\pi \xi} \right) \tag{C.5}$$

$$= \frac{4e^2}{h} \frac{M_{\text{CPS}} \Gamma_{\text{CPS}}^4}{[(\epsilon_{D1} + \epsilon_{D2})^2 + \Gamma_{\text{CPS}}^2]^2}, \tag{C.6}$$

where $k_F$ is the Fermi velocity in the SC and $\xi$ is the SC coherence length. In the second line we have absorbed the geometric suppression and the coupling to the SC in the 'matrix element' $M_{\rm CPS}$. We see that the expression in Eq. (C.6) depends on the sum of the level positions $\epsilon_{D1} + \epsilon_{D2}$.
We can now immediately drop the restriction of $\Gamma_l > \Gamma_{{\rm DS}l}$, $l = 1, 2$, since the relevant tunnel rates are $\Gamma'_{{\rm DS}l} = \Gamma_{{\rm DS}l}\frac{\sin(k_F\delta r)}{k_F\delta r}\exp\left(-\frac{\delta r}{\pi\xi}\right)$, which will be small even if $\Gamma_{{\rm DS}l}$ is not due to the geometric suppression, especially in view of $\delta r \approx 200\,\rm nm$ and $\lambda_F \approx 0.36\,\rm nm$ in a typical experiment (Hofstetter et al. [2009]).
We note that $\Gamma_{\rm CPS} = \Gamma_1 + \Gamma_2$ from the considerations above. The CPS resonance is therefore not affected by the widths $\Gamma_{\rm DS1}$ and $\Gamma_{\rm DS2}$ due to the geometric suppression involved in the relevant quantities $\Gamma'_{\rm DS1}$ and $\Gamma'_{\rm DS2}$.
Next, we discuss the conductance due to EC. The problem is spin-symmetric and we discuss the spin-$\uparrow$ case. The final state for EC from N1 to N2 has the form $|f\rangle = \Psi_{1p\uparrow}\Psi^+_{2q\uparrow}|i\rangle$. Exchanging 1 and 2 in this expression gives the opposite direction for EC which is related to our result by symmetry. We know from our consideration above that the relevant tunnel coupling between the dots via the SC will be small and tunneling can therefore be treated to second order in perturbation theory. Tunneling between the leads and the dot can, however, be resonant and we therefore need to treat tunneling to all orders in $H_{{\rm NT}l}$. Proceeding along the lines of Recher et al. [2001] we can write the transition amplitude between the initial and final state as

$$\langle f|T(\epsilon_i)|i\rangle = \langle \Psi_{1p\uparrow}T'_1 d^+_{1\uparrow}\rangle\langle d_{1\uparrow}d^+_{2\uparrow}T''\rangle\langle d_{2\uparrow}T'_2\Psi^+_{2q\uparrow}\rangle, \tag{C.7}$$

where the partial $T$-matrices $T'_l$ and $T''$ are given by

$$T'' = \frac{1}{i\eta - H_0}H_{\rm ST1}\frac{1}{i\eta - H_0}H_{\rm ST2}\frac{1}{i\eta - H_0}, \text{ and} \tag{C.8}$$

$$T'_l = \sum_{n=0}^{\infty}\left(\frac{1}{i\eta - H_0}H_{{\rm NT}l}\right)^{2n}. \tag{C.9}$$

We first consider the tunnel process in the SC. The emerging integral over the contact area can be taken from Feinberg [2003]. Summing over the electron and hole contribution we obtain

$$\langle d_{1\uparrow}d^+_{2\uparrow}T''\rangle = \frac{\pi\rho_{0s}\gamma_{DS1}\gamma_{DS2}}{(\epsilon_1 - i\eta)(\epsilon_2 - i\eta)}\frac{\cos(k_F\delta r)}{k_F\delta r}e^{-\delta r/\pi\xi}. \tag{C.10}$$

The other two matrix elements in Eq. (C.7) can be taken from Recher et al. [2001]

$$\langle \Psi_{1p\uparrow}T'_1 d^+_{1\uparrow}\rangle = \frac{\epsilon_1 - i\eta}{\epsilon_1 - i\Gamma_1}, \ \langle d_{2\uparrow}T'_2\Psi^+_{2q\uparrow}\rangle = \frac{\epsilon_2 - i\eta}{\epsilon_2 - i\Gamma_2}. \tag{C.11}$$

In conclusion we obtain for the EC conductance

$$G'_{EC} = \frac{4e^2}{h}\frac{4\Gamma_{\rm DS1}\Gamma_{\rm DS2}\Gamma_1\Gamma_2}{(\epsilon_{D1}^2 + \Gamma_1^2)(\epsilon_{D2}^2 + \Gamma_2^2)}\left[\frac{\cos(k_F\delta r)}{k_F\delta r}\right]^2\exp\left(-\frac{2\delta r}{\pi\xi}\right) \tag{C.12}$$

$$= \frac{4e^2}{h}\frac{M_{EC}\Gamma_{\rm EC1}^2\Gamma_{\rm EC2}^2}{(\epsilon_{D1}^2 + \Gamma_{\rm EC1}^2)(\epsilon_{D2}^2 + \Gamma_{\rm EC2}^2)}, \tag{C.13}$$

where we introduced $M_{\rm EC}$ similar to $M_{\rm CPS}$ in Eq. (C.6). Again, we observe the geometric suppression leading to the relevant tunnel couplings $\Gamma''_{\rm DS1,2} = \Gamma_{\rm DS1,2}\frac{\cos(k_F\delta r)}{k_F\delta r}\exp\left(-\frac{\delta r}{\pi\xi}\right)$, so that the result is applicable also for large $\Gamma_{\rm DS1,2}$. From our consideration above we find $\Gamma_{\rm EC1} = \Gamma_1$, $\Gamma_{\rm EC2} = \Gamma_2$.
Finally, we relax the second restriction of small bias $U_{\rm N2} < \Delta/e$. In Section 6.1 we investigated the finite bias case, which basically leads to a replacement of the simple factor $\Gamma_{S1}$ by an energy-dependent effective tunnel rate $\Gamma_{S2}$, see Eq. (6.7). For the nonlocal processes analysed here we

have $\Gamma'_{\mathrm{DS}l} \ll \Gamma_l$, $l = 1, 2$ due to the geometric suppression. In this limit of small transparency, we can use a semiconductor model adapted from Tinkham [ 1996], where we find the finite bias transmission coefficient to be given by the zero bias transmission coefficient multiplied by the DOS of Cooper pairs. In the simplest approximation this one is just a constant below and zero above the SC gap (Kroemer [ 1998]), which amounts to multiplying the zero bias transmission coefficient by a step-function at the SC gap

$$\rho(x) = \theta(\Delta - x)\theta(\Delta + x), \tag{C.14}$$

We compare this approximation to an exact result at the end of this Appendix. It applies since the width of the resonance is basically determined only by the tunnel rates to the normal leads $\Gamma_1$ and $\Gamma_2$ while the SC tunnel rates only contribute to an overall scaling and become ineffective for voltages larger than the gap.

In the following we will use $M_{\mathrm{EC}}$ and $M_{\mathrm{CPS}}$ introduced in Eqs. (C.6) and (C.13) as fitting parameters, since reconciliation of the geometrical suppression in Eq. (C.6) is difficult (Hofstetter et al. [ 2009]), especially in view of many possible effects that could alter the exact form of the suppression.

Finally, we discuss the approximation for finite voltage effects. In Section 6.1 conductance for charge transfer mediated by ARs in a SC-QD-normal metal junction in case of finite bias $V$ and finite temperature has been considered. CPS and EC in the setup considered above are both charge transfer processes below the SC gap with rather low tunneling coupling to the SC due to the geometric suppression. Therefore, a good test for the approximation of the energy dependence in Eq. (C.14) is to compare the approximate conductance resulting from Eq. (C.14) in a SC-QD-normal metal junction at low tunneling coupling to the exact result. For convenience we consider a QD with a resonance at voltage $V = 0$. Eq. (B.2) will give the following result for the conductance (for small $V$ and $T = 0$)

$$G_{\mathrm{NDS,small}\,V}(V) = \frac{4e^2}{h}\left(\frac{2\Gamma_1\Gamma_{S1}}{V^2 + \Gamma_1^2 + \Gamma_{S1}^2}\right)^2. \tag{C.15}$$

If $V \ll \Delta$ is not fulfilled anymore we have to include the abovementioned effects of the energy dependent SC DOS. We obtain the conductance at arbitrary $V$ from the approximation in Eq. (C.14) via

$$G_{\mathrm{NDS}}(V) = G_{\mathrm{NDS,small}\,V}(V)\rho(V). \tag{C.16}$$

We compare the approximation in Eq. (C.16) to the exact result using Eq. (B.2) in Fig. C.1 for typical parameters for the non-local conductances.

We observe acceptable agreement between the exact and approximate solution. They do not agree at $V = 0$ since we perform the calculation at finite temperature as in Eq. (7.12).

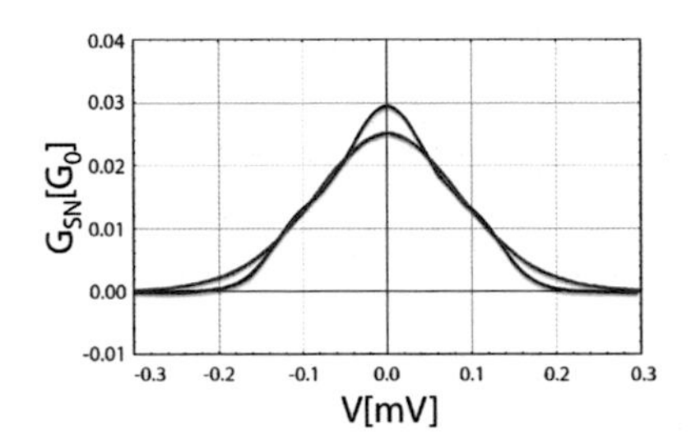


**Figure C.1.:** Bias dependence of the exact conductance (blue curve) using Eq. (B.2) and $G_{\mathrm{NDS,small}\,V}$ (red curve) given by Eq. (C.16) as a function of the applied bias $V$ between the normal metal and the SC. We use parameters $\Gamma_1 = 0.2$ meV, $\Gamma_{S1} = 0.01$ meV, $\Delta = 0.13$ meV and $T = 0.02$ meV in order to represent the typical situation for the non-local conductances. We observe that the approximate solution mostly agrees well with the exact result.

# Appendix D

## Master equation approach to Cooper pair splitting

In the experimental work of Schindele et al. [ 2012] it was suggested to use a master equation approach similar to Eldridge et al. [ 2010] to qualitatively explain the data. Here, we will show how to obtain the necessary rates from the generic model described in Chapter 7 . Indeed, the approach taken in Eldridge et al. [ 2010] cannot be generalized straightforward since the SC was assumed to have an infinite gap which automatically forbids processes such as LPT.
We start by considering only charge eigenstates of the QDs. In this case we do have four states as indicated in Fig. D.1 (a) along with the rates we encounter when the leads N1, N2 have negative bias with respect to the SC. We have not included possible inter-dot processes described by $t_{12}$ in the generic model, since these rates are typically small in the experiment considered (Schindele et al. [2012]). We introduce also a fifth state $(0,1)^*$ highlighted in red which indicates the first part of CPS.

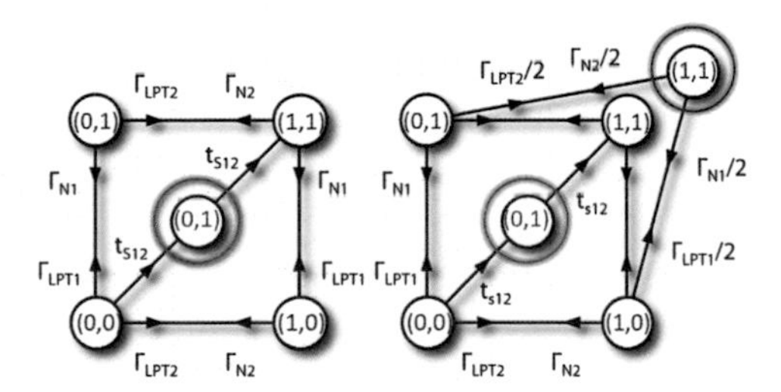


**Figure D.1.:** Sketch of the different states and rates involved in solving the master equation of the carbon nanotube based Cooper pair splitter. **(a)** shows the scheme where only charge eigenstates are considered, whereas **(b)** shows the scheme when we include the difference between a triplet and a singlet state on the two quantum dots.

To lowest order in the rates for CPS, LPT and $\Gamma_1$, $\Gamma_2$ the occupation probabilities obey the master equation

$$\sum_{\chi,\chi'}(W_{\chi\chi'}P_{\chi'} - W_{\chi'\chi}P_{\chi}) = 0, \tag{D.1}$$

where $W_{\chi\chi'}$ is Fermi's golden rule transition rate from state $\chi'$ to $\chi$, where we have $\chi, \chi' = (0,0),\ (0,1),\ (1,0),\ (0,1)^*,\ (1,1)$. In order to simplify notation let us assume zero temperature and the leads and QDs gated in such a way that no backflow to the SC is possible. In this case the non-vanishing rates read

$$\begin{aligned}
W_{(0,0),(0,0)} &= \Gamma_{\mathrm{LPT1}},\ W_{(0,1),(0,0)} = \Gamma_1,\ W_{(0,0),(0,1)^*} = t_{S12},\ W_{(0,1)^*,(1,1)} = t_{S12}, && \text{(D.2)}\\
W_{(0,0),(1,0)} &= \Gamma_{\mathrm{LPT2}},\ W_{(1,0),(0,0)} = \Gamma_2,\ W_{(0,1),(1,1)} = \Gamma_{\mathrm{LPT2}},\ W_{(1,1),(0,1)} = \Gamma_2, && \text{(D.3)}\\
W_{(1,0),(1,1)} &= \Gamma_{\mathrm{LPT1}},\ W_{(1,1),(1,0)} = \Gamma_1, && \text{(D.4)}\\
& && \text{(D.5)}
\end{aligned}$$

The only rates we have not discussed so far are $\Gamma_{\mathrm{LPT}\ 1,2}$. Indeed, LPT is a process which is third order in the tunnel coupling. In order to introduce proper rates we break up each arrow for LPT in Fig. D.1 as indicated in Fig. D.2 .

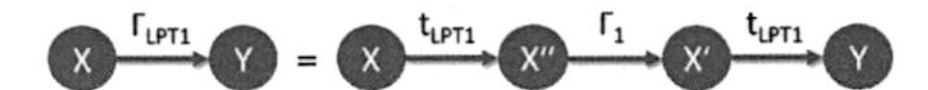


**Figure D.2.:** Sketch of how to write LPT using proper rates. The example is for $\Gamma_{\mathrm{LPT1}}$ and the treatment is analogous for $\Gamma_{\mathrm{LPT2}}$.

From (Recher et al. [ 2001]) we immediately find

$$t_{\mathrm{LPT}} = t_{S12}\frac{(\Gamma_1+\Gamma_2)}{\pi\Delta\left[\frac{\sin(k_F\delta r)}{k_F\delta}\right]\exp\left(-\frac{\delta r}{\pi\xi}\right)}. \tag{D.6}$$

This model now includes the proper theoretical definition of the effective rates $p_{\mathrm{CPS}}$ and $p_{\mathrm{LPT}}$ that were defined in (Schindele et al. [ 2012]).
The current in lead $\eta$ can be computed from the occupation probabilities by

$$I_\eta = \frac{e}{\hbar}\sum_{\chi\neq\chi'} W^{\eta}_{\chi'\chi}P_{\chi}, \tag{D.7}$$

where the current rates $W^{\eta}_{\chi'\chi}$ take into account the electrons transferred to lead $\eta$.
We choose typical parameters $t_{S12} = 0.1 = t_{\mathrm{LPT}}$, $\Gamma_1 = 0.12$, $\Gamma_2 = 0.195$ in units of $\Delta$ and obtain for the total current $I_1 \approx 0.043\Delta e/\hbar$. We can also calculate the CPS current $I_{CPS} = e/\hbar\Gamma_{N1}P_{(1,1)} \approx 0.01\Delta e/\hbar$. However, if we calculate $I_1$ for $t_{S12} = 0$ we obtain $I_1(t_{S12} = 0) = 0.037\Delta e/\hbar \neq I_1 - I_{CPS}$. As discussed in Section 7.3 this is due to the different DOS of the QDs when$t_{S12}$ is present or not. In the master equation this change is reflected by changing probabilities $P_\chi$, which was also found in Schindele et al. [ 2012].
However, two additional complications arise. First, it is not trivial to extract absolute values for the $P_\chi$, especially for the complex processes involving the SC. Secondly, the state $(1,1)$ is not necessarily a singlet state as it can also be populated by two consecutive LPTs with equal spins, which gives rise to a finite triplet amplitude.
In order to demonstrate this possibility we consider the more involved scheme in Fig. D.1 (b),

where we include a triplet state. We implement the master equation as before and calculate the probabilities. We find

$$\frac{P_{(1,1),\text{triplet}}}{P_{(1,1),\text{singlet}}} \approx 0.5, \tag{D.8}$$

so that we conclude that actually a large part is contributed by triplet splitting.

# Bibliography

P. Adroguer, C. Grenier, D. Carpentier, J. Cayssol, P. Degiovanni, and E. Orignac. Probing the helical edge states of a topological insulator by Cooper-pair injection. *Phys. Rev. B*, 82:081303, 2010. doi:10.1103/PhysRevB.82.081303.

A. R. Akhmerov, J. P. Dahlhaus, F. Hassler, M. Wimmer, and C. W. J. Beenakker. Quantized conductance at the Majorana phase transition in a disordered superconducting wire. *Phys. Rev. Lett.*, 106:057001, 2011. doi:10.1103/PhysRevLett.106.057001.

K. F. Albrecht, H. Soller, L. Mühlbacher, and A. Komnik. Transient dynamics and steady state behavior of the Anderson-Holstein model with a superconducting lead. *in preparation*, 2012a.

K. F. Albrecht, H. Wang, L. Mühlbacher, M. Thoss, and A. Komnik. Bistability signatures in nonequilibrium charge transport through molecular quantum dots. *Phys. Rev. B*, 86:081412, 2012b. doi:10.1103/PhysRevB.86.081412.

M. P. Anantram and S. Datta. Current fluctuations in mesoscopic systems with Andreev scattering. *Phys. Rev. B*, 53(24):16390, 1996. doi:10.1103/PhysRevB.53.16390.

G. E. Astrakharchik, J. Boronat, I. L. Kurbakov, and Yu. E. Lozovik. Quantum phase transition in a two-dimensional system of dipoles. *Phys. Rev. Lett.*, 98:060405, 2007. doi:10.1103/PhysRevLett.98.060405.

P. Burset Atienza. *Superconducting proximity effect and nonlocal transport in graphene and carbon nanotubes*. PhD thesis, Universidad Autonoma de Madrid, 2012.

J. Bardeen, L. N. Cooper, and J. R. Schrieffer. Microscopic theory of superconductivity. *Phys. Rev.*, 106(1):162, 1957. doi:10.1103/PhysRev.106.162.

C. W. J. Beenakker. Quantum transport in semiconductor-superconductor microjunctions. *Phys. Rev. B*, 46:12841, 1992. doi:10.1103/PhysRevB.46.12841.

W. Belzig. *Quantum Noise in Mesoscopic Physics*, volume 97 of *NATO Sciences Series II: Mathematica Physics and Chemistry*, chapter Full Counting Statistics of Superconductor–Normal-Metal Heterostructures. Springer Verlag, 2002. URL `http://arxiv.org/abs/cond-mat/0210125`.

W. Belzig and Yu. V. Nazarov. Full current statistics in diffusive normal-superconductor structures. *Phys. Rev. Lett.*, 87(6):067006, 2001a. doi:10.1103/PhysRevLett.87.067006.

W. Belzig and Yu. V. Nazarov. Full counting statistics of electron transfer between superconductors. *Phys. Rev. Lett.*, 87:197006, 2001b. doi:10.1103/PhysRevLett.87.197006.

G. E. Blonder, M. Tinkham, and T. M. Klapwijk. Transition from metallic to tunneling regimes in superconducting microconstrictions: Excess current, charge imbalance, and supercurrent conversion. *Phys. Rev. B*, 25:4515, 1982. doi:10.1103/PhysRevB.25.4515.

J. Börlin, W. Belzig, and C. Bruder. Full counting statistics of a superconducting beam splitter. *Phys. Rev. Lett.*, 88:197001, 2002. doi:10.1103/PhysRevLett.88.197001.

V. Bouchiat, N. Chtchelkatchev, D. Feinberg, G. B. Lesovik, T. Martin, and J. Torrès. Single-walled carbon nanotube-superconductor entangler: noise correlations and Einstein-Podolsky-Rosen states. *Nanotechnol.*, 14(1):77, 2003. doi:10.1088/0957-4484/14/1/318.

A. Braggio, M. Governale, M. G. Pala, and J. König. Superconducting proximity effect in interacting quantum dots revealed by shot noise. *Solid State Comm.*, 151(2):155, 2011. doi:10.1016/j.ssc.2010.10.043.

D. Breyel, T. L. Schmidt, and A. Komnik. Rydberg crystallization detection by statistical means. *Phys. Rev. A*, 86:023405, 2012. doi:10.1103/PhysRevA.86.023405.

D. Breyel, H. Soller, T. L. Schmidt, and A. Komnik. Detecting an exciton crystal by statistical means. *submitted for peer review*, 2013.

A. Brinkman and A. A. Golubov. Crossed Andreev reflection in diffusive contacts: Quasiclassical Keldysh-Usadel formalism. *Phys. Rev. B*, 74:214512, 2006. doi:10.1103/PhysRevB.74.214512.

M. R. Buitelaar, T. Nussbaumer, and C. Schönenberger. Quantum dot in the Kondo regime coupled to superconductors. *Phys. Rev. Lett.*, 89(25):256801, 2002. doi:10.1103/PhysRevLett.89.256801.

P. Burset, W. J. Herrera, and A. Levy Yeyati. Microscopic theory of Cooper pair beam splitters based on carbon nanotubes. *Phys. Rev. B*, 84:115448, 2011. doi:10.1103/PhysRevB.84.115448.

L. V. Butov, A. Zrenner, G. Abstreiter, G. Böhm, and G. Weimann. Condensation of indirect excitons in coupled AlAs/GaAs quantum wells. *Phys. Rev. Lett.*, 73:304, 1994. doi:10.1103/PhysRevLett.73.304.

M. Büttiker, A. Prêtre, and H. Thomas. Dynamic conductance and the scattering matrix of small conductors. *Phys. Rev. Lett.*, 70:4114, Jun 1993. doi:10.1103/PhysRevLett.70.4114.

C. Comte and P. Nozières. Exciton Bose condensation : the ground state of an electron-hole gas - I. Mean field description of a simplified model. *J. Phys. France*, 43(7):1069, 1982. doi:10.1051/jphys:019820043070106900.

C. Caroli, R. Combescot, P. Nozières, and D. Saint-James. Direct calculation of the tunneling current. *J. Phys. C*, 4(8):916, 1971. doi:10.1088/0022-3719/4/8/018.

G. F. Cerofolini. Realistic limits to computation - III. Climbing the third dimension. *Appl. Phys. A*, 106(4):967, 2011. doi:10.1007/s00339-011-6724-2.

D. Chevallier, J. Rech, T. Jonckheere, and T. Martin. Current and noise correlations in a double-dot Cooper-pair beam splitter. *Phys. Rev. B*, 83:125421, 2011. doi:10.1103/PhysRevB.83.125421.

S. B. Chung, H. Zhang, X. Qi, and S. Zhang. Topological superconducting phase and Majorana fermions in half-metal/superconductor heterostructures. *Phys. Rev. B*, 84:060510, 2011. doi:10.1103/PhysRevB.84.060510.

B. S. Cirel'son. Quantum generalizations of Bell's inequality. *Lett. Math. Phys.*, 4:93, 1980. doi:10.1007/BF00417500.

A. A. Clerk, V. Ambegaokar, and S. Hershfield. Andreev scattering and the Kondo effect. *Phys. Rev. B*, 61:3555, 2000. doi:10.1103/PhysRevB.61.3555.

J. M. D. Coey. Materials for spin electronics. In M. Ziese and M. J. Thornton, editors, *Spin Electronics*, volume 569 of *Lecture Notes in Physics*, page 277. Springer Berlin Heidelberg, 2001. URL `http://boulder.research.yale.edu/Boulder-2003/reading/coey_spintronics.pdf`.

M. H. Cohen, L. M. Falicov, and J. C. Phillips. Superconductive tunneling. *Phys. Rev. Lett.*, 8(8): 316, 1962. doi:10.1103/PhysRevLett.8.316.

M. Colci, K. Sun, N. Shah, S. Vishveshwara, and D. J. Van Harlingen. Anomalous polarization-dependent transport in nanoscale double-barrier superconductor/ferromagnet/superconductor junctions. *Phys. Rev. B*, 85:180512, 2012. doi:10.1103/PhysRevB.85.180512.

A. Cottet, B. Douçot, and W. Belzig. Finite frequency noise of a superconductor-ferromagnet quantum point contact. *Phys. Rev. Lett.*, 101(25):257001, 2008. doi:10.1103/PhysRevLett.101.257001.

A. Cottet, D. Huertas-Hernando, W. Belzig, and Yu. V. Nazarov. Spin-dependent boundary conditions for isotropic superconducting Green's functions. *Phys. Rev. B*, 80(18):184511, 2009. doi:10.1103/PhysRevB.80.184511.

S. Csonka, L. Hofstetter, F. Freitag, S. Oberholzer, C. Schönenberger, T. S. Jespersen, M. Aagesen, and J. Nygård. Giant Fluctuations and Gate Control of the g-Factor in InAs Nanowire Quantum Dots. *Nano Lett.*, 8:3932, November 2008. doi:10.1021/nl802418w.

J. C. Cuevas. *Electronic transport in normal and superconducting nanostructures*. PhD thesis, Universidad Autonoma de Madrid, 1999. URL `http://www.uam.es/personal_pdi/ciencias/jcuevas/Publications/thesis.pdf`.

J. C. Cuevas, A. Martín-Rodero, and A. Levy Yeyati. Hamiltonian approach to the transport properties of superconducting quantum point contacts. *Phys. Rev. B*, 54:7366, 1996. doi:10.1103/PhysRevB.54.7366.

J. C. Cuevas, A. Levy Yeyati, and A. Martín-Rodero. Kondo effect in normal-superconductor quantum dots. *Phys. Rev. B*, 63:094515, 2001. doi:10.1103/PhysRevB.63.094515.

J. P. Dahlhaus, S. Maier, and A. Komnik. Spin-polarized current generation and detection by a double quantum dot structure. *Phys. Rev. B*, 81(7):075110, 2010. doi:10.1103/PhysRevB.81.075110.

A. Das, Y. Ronen, M. Heiblum, D. Mahalu, A. V. Kretinin, and H. Shtrikman. High-efficiency Cooper pair splitting demonstrated by two-particle conductance resonance and positive noise cross-correlation. *Nat. Comm.*, 3:1165, 2012. doi:10.1038/ncomms2169.

S. Datta. *Electronic transport in mesoscopic systems*, volume 3. Cambridge University Press, 1997.

S. De Franceschi, L. Kouwenhoven, C. Schönenberger, and W. Wernsdorfer. Hybrid superconductor-quantum dot devices. *Nature Nanotechnol.*, 5:703, 2010. doi:10.1038/nnano.2010.173.

D. J. de Haas and G. J. van den Berg. The electrical resistance of gold and silver at low temperatures. *Physica*, 3:440, 1936. doi:10.1016/S0031-8914(36)80009-3.

A. Di Lorenzo and Yu. V. Nazarov. Full counting statistics with spin-sensitive detectors reveals spin singlets. *Phys. Rev. Lett.*, 94:210601, 2005. doi:10.1103/PhysRevLett.94.210601.

Y. Doh, S. de Franceschi, E. P. A. M. Bakkers, and L. P. Kouwenhoven. Andreev reflection versus coulomb blockade in hybrid semiconductor nanowire devices. *Nano Lett.*, 8(12):4098, 2008. doi:10.1021/nl801454k.

F. Dolcini, D. Rainis, F. Taddei, M. Polini, R. Fazio, and A. H. MacDonald. Blockade and counterflow supercurrent in exciton-condensate Josephson junctions. *Phys. Rev. Lett.*, 104(2):027004, 2010. doi:10.1103/PhysRevLett.104.027004.

M. Duckheim and P. W. Brouwer. Andreev reflection from noncentrosymmetric superconductors and Majorana bound-state generation in half-metallic ferromagnets. *Phys. Rev. B*, 83(5):054513, 2011. doi:10.1103/PhysRevB.83.054513.

R. C. Dynes, V. Narayanamurti, and J. P. Garno. Direct measurement of quasiparticle-lifetime broadening in a strong-coupled superconductor. *Phys. Rev. Lett.*, 41(21):1509, 1978. doi:10.1103/PhysRevLett.41.1509.

R. C. Dynes, J. P. Garno, G. B. Hertel, and T. P. Orlando. Tunneling study of superconductivity near the metal-insulator transition. *Phys. Rev. Lett.*, 53(25):2437, 1984. doi:10.1103/PhysRevLett.53.2437.

U. Eckern, G. Schön, and V. Ambegaokar. Quantum dynamics of a superconducting tunnel junction. *Phys. Rev. B*, 30(11):6419, 1984. doi:10.1103/PhysRevB.30.6419.

A. Einstein, B. Podolsky, and N. Rosen. Can quantum-mechanical description of physical reality be considered complete? *Phys. Rev.*, 47:777, 1935. doi:10.1103/PhysRev.47.777.

J. P. Eisenstein and A. H. MacDonald. Bose-Einstein condensation of excitons in bilayer electron systems. *Nature*, 432:691, 2004. doi:10.1038/nature03081.

J. P. Eisenstein, L. N. Pfeiffer, and K. W. West. Independently contacted two dimensional electron systems in double quantum wells. *Appl. Phys. Lett.*, 57:2324, 1990. doi:10.1063/1.103882.

J. Eldridge, M. G. Pala, M. Governale, and J. König. Superconducting proximity effect in interacting double-dot systems. *Phys. Rev. B*, 82:184507, 2010. doi:10.1103/PhysRevB.82.184507.

G. Falci, D. Feinberg, and F. W. J. Hekking. Correlated tunneling into a superconductor in a multiprobe hybrid structure. *Europhys. Lett.*, 54(2):255, 2001. doi:10.1209/epl/i2001-00303-0.

R. Fazio and R. Raimondi. Resonant Andreev tunneling in strongly interacting quantum dots. *Phys. Rev. Lett.*, 80:2913, 1998. doi:10.1103/PhysRevLett.80.2913.

D. Feinberg. Andreev scattering and cotunneling between two superconductor-normal metal interfaces: the dirty limit. *Eur. Phys. J. B*, 36:419, 2003. doi:10.1140/epjb/e2003-00361-6.

D. K. Ferry and S. M. Goodnick. *Transport in Nanostructures (Cambridge Studies in Semiconductor Physics and Microelectronic Engineering)*. Cambridge University Press, 1997. ISBN 0521461413.

K. Flensberg. Tunneling characteristics of a chain of Majorana bound states. *Phys. Rev. B*, 82(18): 180516, 2010. doi:10.1103/PhysRevB.82.180516.

J. Fransson, O. Eriksson, and I. Sandalov. Effects of non-orthogonality and electron correlations on the time-dependent current through quantum dots. *Phys. Rev. B*, 66:195319, 2002. doi:10.1103/PhysRevB.66.195319.

A. Freyn, M. Flöser, and R. Mélin. Positive current cross-correlations in a highly transparent normal-superconducting beam splitter due to synchronized Andreev and inverse Andreev reflections. *Phys. Rev. B*, 82:014510, 2010. doi:10.1103/PhysRevB.82.014510.

A. Freyn, B. Douçot, D. Feinberg, and R. Mélin. Production of nonlocal quartets and phase-sensitive entanglement in a superconducting beam splitter. *Phys. Rev. Lett.*, 106:257005, 2011. doi:10.1103/PhysRevLett.106.257005.

S. Gangadharaiah, B. Braunecker, P. Simon, and D. Loss. Majorana edge states in interacting one-dimensional systems. *Phys. Rev. Lett.*, 107:036801, Jul 2011. doi:10.1103/PhysRevLett.107.036801.

A. Gärtner, L. Prechtel, D. Schuh, A. W. Holleitner, and J. P. Kotthaus. Micropatterned electrostatic traps for indirect excitons in coupled GaAs quantum wells. *Phys. Rev. B*, 76:085304, 2007. doi:10.1103/PhysRevB.76.085304.

I. Giaever and K. Megerle. Study of superconductors by electron tunneling. *Phys. Rev.*, 122:1101, 1961. doi:10.1103/PhysRev.122.1101.

A. O. Gogolin and A. Komnik. Full counting statistics for the Kondo dot in the unitary limit. *Phys. Rev. Lett.*, 97:016602, 2006a. doi:10.1103/PhysRevLett.97.016602.

A. O. Gogolin and A. Komnik. Towards full counting statistics for the Anderson impurity model. *Phys. Rev. B*, 73:195301, 2006b. doi:10.1103/PhysRevB.73.195301.

D. Goldhaber-Gordon, H. Shtrikman, D. Mahalu, D. Abusch-Magder, U. Meirav, and M. A. Kastner. Kondo effect in a single-electron transistor. *Nature*, 391:156, 1998. doi:10.1038/34373.

D. S. Golubev and A. D. Zaikin. Shot noise and Coulomb effects on nonlocal electron transport in normal-metal/superconductor/normal-metal heterostructures. *Phys. Rev. B*, 82(13):134508, 2010. doi:10.1103/PhysRevB.82.134508.

R. V. Gorbachev, A. K. Geim, M. I. Katsnelson, K. S. Novoselov, T. Tudorovskiy, I. V. Grigorieva, A. H. MacDonald, K. Watanabe, T. Taniguchi, and L. A. Ponomarenko. Strong Coulomb drag and broken symmetry in double-layer graphene. *Nat. Phys.*, 8:896, 2012. doi:10.1038/nphys2441.

M.R. Gräber, T. Nussbaumer, W. Belzig, and C. Schönenberger. Quantum dot coupled to a normal and a superconducting lead. *Nanotechnol.*, 15(7):479, 2004. doi:10.1088/0957-4484/15/7/056.

R. Grein, M. Eschrig, G. Metalidis, and Gerd Schön. Spin-dependent Cooper pair phase and pure spin supercurrents in strongly polarized ferromagnets. *Phys. Rev. Lett.*, 102(22):227005, 2009. doi:10.1103/PhysRevLett.102.227005.

R. Grein, T. Löfwander, G. Metalidis, and M. Eschrig. Theory of superconductor-ferromagnet point-contact spectra: The case of strong spin polarization. *Phys. Rev. B*, 81(9):094508, 2010. doi:10.1103/PhysRevB.81.094508.

S. Gustavsson, R. Leturcq, T. Ihn, K. Ensslin, M. Reinwald, and W. Wegscheider. Measurements of higher-order noise correlations in a quantum dot with a finite bandwidth detector. *Phys. Rev. B*, 75:075314, 2007. doi:10.1103/PhysRevB.75.075314.

D. R. Hamann. Orthogonality catastrophe in metals. *Phys. Rev. Lett.*, 26:1030, 1971. doi:10.1103/PhysRevLett.26.1030.

W. J. Herrera, P. Burset, and A. Levy Yeyati. A Green function approach to graphene-superconductor junctions with well-defined edges. *J. Phys.: Condens. Matt.*, 22(27):275304, 2010. doi:10.1088/0953-8984/22/27/275304.

L. G. Herrmann, F. Portier, P. Roche, A. Levy Yeyati, T. Kontos, and C. Strunk. Carbon nanotubes as Cooper-pair beam splitters. *Phys. Rev. Lett.*, 104:026801, 2010. doi:10.1103/PhysRevLett.104.026801.

L. G. Herrmann, P. Burset, W. J. Herrera, F. Portier, P. Roche, C. Strunk, A. Levy Yeyati, and T. Kontos. Spectroscopy of non-local superconducting correlations in a double quantum dot. *ArXiv e-prints*, 2012. URL `http://arxiv.org/abs/1205.1972`.

A. A. High, J. R. Leonard, A. T. Hammack, M. M. Fogler, L. V. Butov, A. V. Kavokin, K. L. Campman, and A. C. Gossard. Spontaneous coherence in a cold exciton gas. *Nature*, 483:584, 2012. doi:10.1038/nature10903.

L. Hofstetter. *Hybrid quantum dots in InAs nanowires*. PhD thesis, Universität Basel, 2011. URL `http://www.nanoelectronics.ch/publications/theses/2011_Thesis_Lukas_Hofstetter.pdf`.

L. Hofstetter, S. Csonka, J. Nygård, and C. Schönenberger. Cooper pair splitter realized in a two-quantum-dot Y-junction. *Nature*, 461:960, 2009. doi:10.1038/nature08432.

L. Hofstetter, A. Geresdi, M. Aagesen, J. Nygård, C. Schönenberger, and S. Csonka. Ferromagnetic proximity effect in a ferromagnet–quantum-dot–superconductor device. *Phys. Rev. Lett.*, 104: 246804, 2010. doi:10.1103/PhysRevLett.104.246804.

L. Hofstetter, S. Csonka, A. Baumgartner, G. Fülöp, S. d'Hollosy, J. Nygård, and C. Schönenberger. Finite-bias Cooper pair splitting. *Phys. Rev. Lett.*, 107:136801, 2011. doi:10.1103/PhysRevLett.107.136801.

F. Hübler, M. J. Wolf, T. Scherer, D. Wang, D. Beckmann, and H. v. Löhneysen. Observation of Andreev bound states at spin-active interfaces. *Phys. Rev. Lett.*, 109:087004, 2012. doi:10.1103/PhysRevLett.109.087004.

A. Jauho, N. S. Wingreen, and Y. Meir. Time-dependent transport in interacting and noninteracting resonant-tunneling systems. *Phys. Rev. B*, 50:5528, 1994. doi:10.1103/PhysRevB.50.5528.

X. Jehl, M. Sanquer, R. Calemczuk, and D. Mailly. Detection of doubled shot noise in short normal-metal/superconductor junctions. *Nature*, 405:50, 2000. doi:10.1038/35011012.

J. Johansson, K. A. Dick, P. Caroff, M. E. Messing, J. Bolinsson, K. Deppert, and L. Samuelson. Diameter dependence of the wurtzite-zinc blende transition in InAs nanowires. *J. Phys. Chem. C*, 114(9):3837, 2010. doi:10.1021/jp910821e.

J. B. Johnson. Thermal agitation of electricity in conductors. *Phys. Rev.*, 32(1):97, Jul 1928. doi:10.1103/PhysRev.32.97.

K. Joho, S. Maier, and A. Komnik. Transient noise spectra in resonant tunneling setups: Exactly solvable models. *Phys. Rev. B*, 86:155304, 2012. doi:10.1103/PhysRevB.86.155304.

T. Jonckheere, A. Zazunov, K. V. Bayandin, V. Shumeiko, and T. Martin. Nonequilibrium supercurrent through a quantum dot: Current harmonics and proximity effect due to a normal-metal lead. *Phys. Rev. B*, 80:184510, 2009. doi:10.1103/PhysRevB.80.184510.

D. Kast, L. Kecke, and J. Ankerhold. Charge transfer through single molecule contacts: How reliable are rate descriptions? *Beilstein J. Nanotech.*, 2:416, 2011. doi:10.3762/bjnano.2.47.

A. Yu. Kitaev. Unpaired Majorana fermions in quantum wires. *Phys. Uspekhi*, 44(10):131, 2001. doi:10.1070/1063-7869/44/10S/S29.

A. Komnik and H. Saleur. Full counting statistics of chiral Luttinger liquids with impurities. *Phys. Rev. Lett.*, 96:216406, 2006. doi:10.1103/PhysRevLett.96.216406.

J. Kondo. Reistance minimum in dilute magnetic alloys. *Prog. Theor. Phys.*, 32:37, 1964. doi:10.1143/PTP.32.37.

J. König, H. Schoeller, and G. Schön. Zero-bias anomalies and boson-assisted tunneling through quantum dots. *Phys. Rev. Lett.*, 76:1715, 1996. doi:10.1103/PhysRevLett.76.1715.

L. P. Kouwenhoven and L. Glazman. Revival of the Kondo effect. *Phys. World*, 14(1):33, 2001. URL `http://arxiv.org/abs/cond-mat/0104100`.

K. Kowalik-Seidl, X. P. Vogele, B. N. Rimpfl, S. Manus, J. P. Kotthaus, D. Schuh, W. Wegscheider, and A. W. Holleitner. Long exciton spin relaxation in coupled quantum wells. *Appl. Phys. Lett.*, 97(1):011104, 2010. doi:10.1063/1.3458703.

H. Kroemer. Supercurrent flow through a semiconductor: the transport properties of superconductor-semiconductor hybrid structures. *Physica E*, 2:887, 1998. doi:10.1016/S1386-9477(98)00181-7.

H. Kubota, A. Fukushima, K. Yakushiji, T. Nagahama, S. Yuasa, K. Ando, H. Maehara, Y. Nagamine, K. Tsunekawa, D. D. Djayaprawira, N. Watanabe, and Y. Suzuki. Quantitative measurement of voltage dependence of spin-transfer torque in MgO-based magnetic tunnel junctions. *Nat. Phys.*, 4:37, 2008. doi:10.1038/nphys784.

D. Kulakovskii, Yu. Lozovik, and A. Chaplik. Collective excitations in exciton crystal. *JETP*, 99: 850, 2004. doi:10.1134/1.1826178.

C. J. Lambert. Generalized Landauer formula for quasi-particle transport in disordered superconductors. *J. Phys.: Condens. Matter*, 3:6579, 1991. doi:10.1088/0953-8984/3/34/003.

D. C. Langreth. *Linear and nonlinear electron transport in solids*, chapter Linear and Non-Linear Response Theory with applications. Plenum Press, 1976.

A. J. Leggett, S. Chakravarty, A. T. Dorsey, Matthew P. A. Fisher, A. Garg, and W. Zwerger. Dynamics of the dissipative two-state system. *Rev. Mod. Phys.*, 59:1, 1987. doi:10.1103/RevModPhys.59.1.

G. B. Lesovik, T. Martin, and G. Blatter. Electronic entanglement in the vicinity of a superconductor. *Eur. Phys. J. B*, 24(3):287, 2001. doi:10.1007/s10051-001-8675-4.

R. Leturcq, C. Stampfer, K. Inderbitzin, L. Durrer, C. Hierold, E. Mariani, M. G. Schultz, F. von Oppen, and K. Ensslin. Franck-Condon blockade in suspended carbon nanotube quantum dots. *Nat. Phys.*, 5:327, 2009. doi:10.1038/nphys1234.

L. S. Levitov and G. B. Lesovik. Quantum Measurement in Electric Circuit. *ArXiv e-prints*, 1994. URL `http://arxiv.org/abs/cond-mat/9401004`.

L. S. Levitov and M. Reznikov. Counting statistics of tunneling current. *Phys. Rev. B*, 70(11): 115305, 2004. doi:10.1103/PhysRevB.70.115305.

W. Liang, M. P. Shores, M. Bockrath, J. R. Long, and H. Park. Kondo resonance in a single-molecule transistor. *Nature*, 417:725, 2002. doi:10.1038/nature00790.

M. Lindberg and S. W. Koch. Effective Bloch equations for semiconductors. *Phys. Rev. B*, 38:3342, 1988. doi:10.1103/PhysRevB.38.3342.

T. Löfwander, R. Grein, and M. Eschrig. Is $CrO_2$ fully spin polarized? Analysis of Andreev spectra and excess current. *Phys. Rev. Lett.*, 105:207001, 2010. doi:10.1103/PhysRevLett.105.207001.

D. Loss and D. P. DiVincenzo. Quantum computation with quantum dots. *Phys. Rev. A*, 57:120, 1998. doi:10.1103/PhysRevA.57.120.

Yu. E. Lozovik and O. L. Berman. Phase transitions in a system of spatially separated electrons and holes. *Zh. Éksp. Teor. Fiz. [JETP 84, 1027 (1997)]*, 111:1879, 1997. doi:10.1134/1.558220.

Yu. E. Lozovik, I. V. Ovchinnikov, S. Yu. Volkov, L. V. Butov, and D. S. Chemla. Quasi-two-dimensional excitons in finite magnetic fields. *Phys. Rev. B*, 65:235304, 2002. doi:10.1103/PhysRevB.65.235304.

S. Maier. *Effect of electron-phonon interaction in nanostructures and ultracold quantum gases.* PhD thesis, Ruprecht-Karls Universität Heidelberg, 2011. URL `http://www.ub.uni-heidelberg.de/archiv/12839`.

S. Maier, T. L. Schmidt, and A. Komnik. Charge transfer statistics of a molecular quantum dot with strong electron-phonon interaction. *Phys. Rev. B*, 83:085401, 2011. doi:10.1103/PhysRevB.83.085401.

A. Martín-Rodero, A. Levy Yeyati, and J.C. Cuevas. Transport properties of normal and ferromagnetic atomic-size constrictions with superconducting electrodes. *Physica C*, 352(1-4):67, 2001. doi:10.1016/S0921-4534(00)01679-8.

J. Martinek, M. Sindel, L. Borda, J. Barnaś, R. Bulla, J. König, G. Schön, S. Maekawa, and J. von Delft. Gate-controlled spin splitting in quantum dots with ferromagnetic leads in the Kondo regime. *Phys. Rev. B*, 72(12):121302, 2005. doi:10.1103/PhysRevB.72.121302.

R. Mélin. Microscopic theory of equilibrium properties of F/S/F trilayers with weak ferromagnets. *Eur. Phys. J. B*, 39:249, 2004. doi:10.1140/epjb/e2004-00188-7.

Z. Merali. Entanglement on demand. *Nature*, 2009. doi:10.1038/news.2009.1002.

H. Min, R. Bistritzer, J. J. Su, and A. H. MacDonald. Room-temperature superfluidity in graphene bilayers. *Phys. Rev. B*, 78(12):121401, Sep 2008. doi:10.1103/PhysRevB.78.121401.

J. P. Morten, D. Huertas-Hernando, W. Belzig, and A. Brataas. Elementary charge transfer processes in a superconductor-ferromagnet entangler. *Europhys. Lett.*, 81(4):40002, 2008. doi:10.1209/0295-5075/81/40002.

V. Mourik, K. Zuo, S. M. Frolov, S. R. Plissard, E. P. A. M. Bakkers, and L. P. Kouwenhoven. Signatures of Majorana fermions in hybrid superconductor-semiconductor nanowire devices. *Science*, 336(6084):1003, 2012. doi:10.1126/science.1222360.

C. Mudry, P. W. Brouwer, and A. Furusaki. Crossover from the chiral to the standard universality classes in the conductance of a quantum wire with random hopping only. *Phys. Rev. B*, 62(12): 8249, 2000. doi:10.1103/PhysRevB.62.8249.

L. Mühlbacher and E. Rabani. Real-time path integral approach to nonequilibrium many-body quantum systems. *Phys. Rev. Lett.*, 100:176403, 2008. doi:10.1103/PhysRevLett.100.176403.

L. Mühlbacher, D. F. Urban, and A. Komnik. Anderson impurity model in nonequilibrium: Analytical results versus quantum Monte Carlo data. *Phys. Rev. B*, 83:075107, 2011. doi:10.1103/PhysRevB.83.075107.

H. Murakawa, K. Ishida, K. Kitagawa, Z. Q. Mao, and Y. Maeno. Measurement of the $^{101}$Ru-knight shift of superconducting $Sr_2RuO_4$ in a parallel magnetic field. *Phys. Rev. Lett.*, 93:167004, 2004. doi:10.1103/PhysRevLett.93.167004.

B. A. Muzykantskii and D. E. Khmelnitskii. Quantum shot noise in a normal-metal–superconductor point contact. *Phys. Rev. B*, 50(6):3982, 1994. doi:10.1103/PhysRevB.50.3982.

D. Nandi, A. D. K. Finck, J. P. Eisenstein, L. N. Pfeiffer, and K. W. West. Exciton Condensation and Perfect Coulomb Drag. *Nature*, 488:481, 2012. doi:10.1038/nature11302.

Yu. V. Nazarov. Universalities of weak localisation. *Ann. Phys.*, 8:193, 1999. URL `http://arxiv.org/abs/cond-mat/9908143`.

Yu. V. Nazarov and M. Kindermann. Full counting statistics of a general quantum mechanical variable. *Eur. Phys. J. B*, 35:413, 2003. doi:10.1140/epjb/e2003-00293-1.

T. K. Ng and P. A. Lee. On-site Coulomb repulsion and resonant tunneling. *Phys. Rev. Lett.*, 61: 1768, 1988. doi:10.1103/PhysRevLett.61.1768.

M. A. Nielsen and I. L. Chuang. *Quantum Computation and quantum Information*. Cambridge University Press, 2000.

P. Nordlander, M. Pustilnik, Y. Meir, N. S. Wingreen, and D. C. Langreth. How long does it take for the Kondo effect to develop? *Phys. Rev. Lett.*, 83:808, 1999. doi:10.1103/PhysRevLett.83.808.

C. Nozières, P. Comte. Exciton Bose condensation : the ground state of an electron-hole gas - II. Spin states, screening and band structure effects. *J. Phys. France*, 43(7):1083, 1982. doi:10.1051/jphys:019820043070108300.

P. Nozières. A "fermi-liquid" description of the Kondo problem at low temperatures. *J. Low Temp. Phys.*, 17:31, 1974. doi:10.1007/BF00654541.

H. Nyquist. Thermal agitation of electric charge in conductors. *Phys. Rev.*, 32(1):110, Jul 1928. doi:10.1103/PhysRev.32.110.

Y. Oreg, G. Refael, and F. von Oppen. Helical liquids and Majorana bound states in quantum wires. *Phys. Rev. Lett.*, 105:177002, 2010. doi:10.1103/PhysRevLett.105.177002.

T. Östreich and A. Knorr. Various appearances of Rabi oscillations for $2\pi$-pulse excitation in a semiconductor. *Phys. Rev. B*, 48:17811, 1993. doi:10.1103/PhysRevB.48.17811.

J. Paaske and K. Flensberg. Vibrational sidebands and the Kondo effect in molecular transistors. *Phys. Rev. Lett.*, 94:176801, 2005. doi:10.1103/PhysRevLett.94.176801.

F. Pérez-Willard, J. C. Cuevas, C. Sürgers, P. Pfundstein, J. Kopu, M. Eschrig, and H. v. Löhneysen. Determining the current polarization in Al/Co nanostructured point contacts. *Phys. Rev. B*, 69: 140502, 2004. doi:10.1103/PhysRevB.69.140502.

S. R. Plissard, D. R. Slapak, M. A. Verheijen, M. Hocevar, G. W. G. Immink, I. van Weperen,

S. Nadj-Perge, S. M. Frolov, L. P. Kouwenhoven, and E. P. A. M. Bakkers. From InSb nanowires to nanocubes: Looking for the sweet spot. *Nano Lett.*, 12(4):1794, 2012. doi:10.1021/nl203846g.

J. Rech, D. Chevallier, T. Jonckheere, and T. Martin. Current correlations in an interacting Cooper-pair beam splitter. *Phys. Rev. B*, 85:035419, 2012. doi:10.1103/PhysRevB.85.035419.

P. Recher, E. V. Sukhorukov, and D. Loss. Andreev tunneling, Coulomb blockade, and resonant transport of nonlocal spin-entangled electrons. *Phys. Rev. B*, 63(16):165314, 2001. doi:10.1103/PhysRevB.63.165314.

B. Reulet, J. Senzier, and D. E. Prober. Environmental effects in the third moment of voltage fluctuations in a tunnel junction. *Phys. Rev. Lett.*, 91:196601, 2003. doi:10.1103/PhysRevLett.91.196601.

F. Robicheaux and J. V. Hernández. Many-body wave function in a dipole blockade configuration. *Phys. Rev. A*, 72:063403, 2005. doi:10.1103/PhysRevA.72.063403.

S. M. Roy. Multipartite separability inequalities exponentially stronger than local reality inequalities. *Phys. Rev. Lett.*, 94:010402, 2005. doi:10.1103/PhysRevLett.94.010402.

J. Schindele, A. Baumgartner, and C. Schönenberger. Near-unity cooper pair splitting efficiency. *Phys. Rev. Lett.*, 109:157002, 2012. doi:10.1103/PhysRevLett.109.157002.

T. L. Schmidt. *Correlation effects in the charge and spin transport statistics of quantum impurity models*. PhD thesis, Albert-Ludwigs Universität Freiburg im Breisgau, 2007. URL `http://www.freidok.uni-freiburg.de/volltexte/3509/`.

T. L. Schmidt, A. Komnik, and A. O. Gogolin. Full counting statistics of spin transfer through ultrasmall quantum dots. *Phys. Rev. B*, 76:241307, 2007a. doi:10.1103/PhysRevB.76.241307.

T. L. Schmidt, A. Komnik, and A. O. Gogolin. Hanbury Brown–Twiss correlations and noise in the charge transfer statistics through a multiterminal Kondo dot. *Phys. Rev. Lett.*, 98:056603, 2007b. doi:10.1103/PhysRevLett.98.056603.

T. L. Schmidt, P. Werner, L. Mühlbacher, and A. Komnik. Transient dynamics of the anderson impurity model out of equilibrium. *Phys. Rev. B*, 78:235110, 2008. doi:10.1103/PhysRevB.78.235110.

C. Schönenberger. Eine Trenneinrichtung für Quantenpaare. *Physik in unserer Zeit*, 41(2):58, 2010. doi:10.1002/piuz.201090007.

W. Schottky. Über spontane Stromschwankungen in verschiedenen Elektrizitätsleitern. *Ann. Phys.*, 362:541, 1918. doi:10.1002/andp.19183622304.

P. Schwab and R. Raimondi. Andreev tunneling in quantum dots: A slave-boson approach. *Phys. Rev. B*, 59:1637, 1999. doi:10.1103/PhysRevB.59.1637.

B. Seradjeh. Majorana edge modes of topological exciton condensate with superconductors. *Phys. Rev. B*, 86:121101, 2012. doi:10.1103/PhysRevB.86.121101.

N. Shah and A. Rosch. Nonequilibrium conductance of a three-terminal quantum dot in the

Kondo regime: Perturbative renormalization group study. *Phys. Rev. B*, 73:081309, 2006. doi:10.1103/PhysRevB.73.081309.

M. Sindel, L. Borda, J. Martinek, R. Bulla, J. König, G. Schön, S. Maekawa, and J. von Delft. Kondo quantum dot coupled to ferromagnetic leads: Numerical renormalization group study. *Phys. Rev. B*, 76:045321, 2007. doi:10.1103/PhysRevB.76.045321.

D. Snoke. Spontaneous Bose coherence of excitons and polaritons. *Science*, 298(5597):1368, 2002. doi:10.1126/science.1078082.

D. Snoke, S. Denev, Y. Liu, L. Pfeiffer, and K. West. Long-range transport in excitonic dark states in coupled quantum wells. *Nature*, 418:754, 2002. doi:10.1038/nature00940.

H. Soller. Full counting statistics for superconducting quantum point contacts. Master's thesis, Ruprecht-Karls Universität Heidelberg, 2009.

H. Soller. Fcs of superconducting tunnel junctions in non-equilibrium. *Internat. J. Mod. Phys. B*, 27, 2013. URL `http://arxiv.org/abs/1302.1106`.

H. Soller and D. Breyel. Detecting phase transitions in excitonic systems via conductance. *in preparation*, 2013.

H. Soller and A. Komnik. Hamiltonian approach to the charge transfer statistics of Kondo quantum dots contacted by a normal metal and a superconductor. *Physica E*, 44(2):425, 2011a. doi:10.1016/j.physe.2011.09.014.

H. Soller and A. Komnik. Charge transfer statistics and entanglement in normal-quantum dot-superconductor hybrid structures. *Eur. Phys. J. D*, 63(1):3, 2011b. doi:10.1140/epjd/e2010-00256-7.

H. Soller and A. Komnik. P-wave Cooper pair splitting. *Beilstein J. Nanotechnol.*, 3:493, 2012. doi:10.3762/bjnano.3.56.

H. Soller and A. Komnik. Current noise and higher order fluctuations in semiconducting bilayer systems. *Fluctuations and Noise Letters*, 12(2), 2013.

H. Soller and A. Wedemeier. Prediction of synergistic multi-compound mixtures - a generalized Colby approach. *Crop Protection*, 42:180, 2012.

H. Soller, P. Burset, L. Hofstetter, A. Baumgartner, B. Braunecker, K. Kang, C. Schönenberger, A. Komnik, and A. Levy Yeyati. The generic model of Cooper pair splitting. *in preparation*, 2012a.

H. Soller, J. P. Dahlhaus, and A. Komnik. Charge transfer statistics of transport through Majorana bound states. *in preparation*, 2012b.

H. Soller, F. Dolcini, and A. Komnik. Nanotransformation and current fluctuations in exciton condensate junctions. *Phys. Rev. Lett.*, 108:156401, 2012c. doi:10.1103/PhysRevLett.108.156401.

H. Soller, L. Hofstetter, S. Csonka, A. Levy Yeyati, C. Schönenberger, and A. Komnik. Kondo effect and spin-active scattering in ferromagnet-superconductor junctions. *Phys. Rev. B*, 85:174512, 2012d. doi:10.1103/PhysRevB.85.174512.

H. Soller, L. Hofstetter, and D. Reeb. Entanglement witnessing in superconducting beamsplitters. *submitted for peer review*, 2012e.

H. Soller, F. Kulmann, and W. Rödder. Efficient and exact solutions for job shop scheduling problems via a generalized colouring algorithm for tree-like graphs. *submitted for peer review*, 2012f.

Henning Soller. Hg-lg mode conversion with stressed 3-mode fibers under polarization. *Open J. Appl. Sci.*, 2(4):224, 2012. doi:10.4236/ojapps.2012.24033.

R. J. Soulen, J. M. Byers, M. S. Osofsky, B. Nadgorny, T. Ambrose, S. F. Cheng, P. R. Broussard, C. T. Tanaka, J. Nowak, J. S. Moodera, A. Barry, and J. M. D. Coey. Measuring the spin polarization of a metal with a superconducting point contact. *Science*, 282(5386):85, 1998. doi:10.1126/science.282.5386.85.

J. J. Su and A. H. MacDonald. How to make a bilayer exciton condensate flow. *Nat. Phys.*, 4:799, 2008. doi:10.1038/nphys1055.

D. J. Thouless. Strong-coupling limit in the theory of Superconductivity. *Phys. Rev.*, 117:1256, 1960. doi:10.1103/PhysRev.117.1256.

M. Tinkham. *Introduction to Superconductivity.* McGraw Hill, 1996.

A.M. Tsvelick and P.B. Wiegmann. Exact results in the theory of magnetic alloys. *Adv. Phys.*, 32 (4):453, 1983. doi:10.1080/00018738300101581.

H. E. Türeci, M. Hanl, M. Claassen, A. Weichselbaum, T. Hecht, B. Braunecker, A. Govorov, L. Glazman, A. Imamoglu, and J. von Delft. Many-body dynamics of exciton creation in a quantum dot by optical absorption: A quantum quench towards Kondo correlations. *Phys. Rev. Lett.*, 106:107402, 2011. doi:10.1103/PhysRevLett.106.107402.

J. Uffink and M. Seevinck. Strengthened Bell inequalities for orthogonal spin directions. *Phys. Lett. A*, 372(8):1205, 2008. doi:10.1016/j.physleta.2007.09.033.

E. Vecino, A. Martín-Rodero, and A. Levy Yeyati. Josephson current through a correlated quantum level: Andreev states and pi junction behavior. *Phys. Rev. B*, 68:035105, 2003. doi:10.1103/PhysRevB.68.035105.

A.F. Volkov, P.H.C. Magnée, B.J. van Wees, and T.M. Klapwijk. Proximity and Josephson effects in superconductor-two-dimensional electron gas planar junctions. *Physica C*, 242(3-4):261, 1995. doi:10.1016/0921-4534(94)02429-4.

Z. Vörös, R. Balili, D. W. Snoke, L. Pfeiffer, and K. West. Long-distance diffusion of excitons in double quantum well structures. *Phys. Rev. Lett.*, 94:226401, 2005. doi:10.1103/PhysRevLett.94.226401.

C. Wang, Y. T. Cui, J. A. Katine, R. A. Buhrman, and D. C. Ralph. Time-resolved measurement of spin-transfer-driven ferromagnetic resonance and spin torque in magnetic tunnel junctions. *Nat. Phys.*, 7:496, 2011. doi:10.1038/nphys1928.

R. F. Werner. Quantum states with Einstein-Podolsky-Rosen correlations admitting a hidden-variable model. *Phys. Rev. A*, 40:4277, 1989. doi:10.1103/PhysRevA.40.4277.

W. Wiethege, P. Entel, and B. Mühlschlegel. Magnetism and superconductivity in the extended periodic Anderson model. *Z. Phys. B*, 47:35, 1982. doi:10.1007/BF01686181.

K. Xia, P. J. Kelly, G. E. W. Bauer, and I. Turek. Spin-dependent transparency of ferromagnet/superconductor interfaces. *Phys. Rev. Lett.*, 89(16):166603, Sep 2002. doi:10.1103/PhysRevLett.89.166603.

Y. Yamada, Y. Tanaka, and N. Kawakami. Magnetic field effects on the Andreev reflection through a quantum dot. *Physica E*, 40:265, 2007. doi:10.1016/j.physe.2007.06.010.

Y. Yamada, Y. Tanaka, and N. Kawakami. Interplay of Kondo and superconducting correlations in the nonequilibrium Andreev transport through a quantum dot. *Phys. Rev. B*, 84:075484, 2011. doi:10.1103/PhysRevB.84.075484.

K. Yosida and K. Yamada. Perturbation expansion for the Anderson Hamiltonian (3). *Prog. Theor. Phys.*, 53:1286, 1975. doi:10.1143/PTP.53.1286.

M. W. Zwierlein, C. A. Stan, C. H. Schunck, S. M. F. Raupach, A. J. Kerman, and W. Ketterle. Condensation of pairs of fermionic atoms near a Feshbach resonance. *Phys. Rev. Lett.*, 92:120403, 2004. doi:10.1103/PhysRevLett.92.120403.

# Acknowledgement

> We have been told we cannot do this by a chorus of cynics. And they will only grow louder and more dissonant in the weeks and months to come. We've been asked to pause for a reality check. We've been warned against offering the people of this nation false hope. But in the unlikely story that is America, there has never been anything false about hope. For when we have faced down impossible odds, when we've been told we're not ready or that we shouldn't try or that we can't, generations of Americans have responded with a simple creed that sums up the spirit of a people:
> Yes, we can.
>
> *(Barack Obama, 2008)*

First I would like to thank Andreas Komnik for giving me the opportunity to do a PhD in his group. He shared his deep insights into the world of physics with me and working with him was always challenging. I also like to thank him for the confidence he gave to me when I was following my own interests. I would also like to thank my (almost) co-advisor Alfredo Levy Yeyati who introduced me to the world of superconductors and was always willing to discuss my ideas and help me with his great intuition for physics.
I would also like to thank Jörg Evers for being the second referee of this thesis and Norbert Herrmann and Matthias Bartelmann for being the additional referees. It is a great honor to have these people taking their time to read my thesis.
However, this work would have been impossible without a lot of people sharing their ideas with me and taking the time for numerous discussions. Lukas Hofstetter provided me with invaluable insights into swiss engineering of InAs nanowires and has been a constant source of cool ideas on how to proceed. He is also a cool guy, by the way. Fabrizio Dolcini has invested a lot of time and effort into convincing us of the importance of exciton condensates and his constant drive for perfection have greatly improved my understanding. I very much appreciated working to together with Ferdi Albrecht in the 'most unprofessional collaboration ever' and thank him for numerous discussions on physics and, of course, also everything else. I am also very grateful to Pablo Burset who shared his great experience with superconductor-graphene interfaces with me and working with him on Cooper pair splitting was great fun. Besides, he also introduced me to Madrid nightlife. I deeply appreciate the help of my very good friend David Reeb, not only when trying to understand entanglement but also for every other problem in physics. His deep understanding is often just admirable. Jan Dahlhaus was always willing to discuss my ideas for the Majoranas and contributed several new ideas on how to proceed. The same is true for Bernd Braunecker who developed into my personal dictionary on every kind of method. Thomas Schmidt has always been an inspiring

source of ideas. Guys, thanks a lot.
I am also very greatly indebted to my colleagues in my research group including David Breyel, Stefan Maier and Karsten Joho. David has taken a lot of time and effort to investigate the exciton crystals with me and joined me on numerous conferences and even went to the strangest bars. Stefan has not only taken a lot of time for discussions and explaining proper physics and writing to me but also made it to the weights room with me. Karsten was very helpful when discussing transient phenomena. Besides, they have all helped me when writing this thesis and did a great job on proof-reading which makes me heavily indebted to them. Thank you all!
I am also particularly grateful for the unique atmosphere at Philosophenweg 19 and special thanks go to my friends and colleagues Giulio Schober, Kai Giering, Kambis Veschgini, Long Lu, Mischa Gerstenlauer, Christoph Orth, Philipp Albert and, of course, Frank Hantschel for sharing with me one of the greatest times in my life... and surely numerous hours in front of my beloved coffee machine. Thanks a lot.
I do not want to forget my friends from the heavy-ion group Yapeng Zhang and Pierre Loizeau... and surely most of all Ingo Deppner, who provided me both with his experimental physics point of view and an unlimited dictionary of new excuses for not starting to write a thesis or coming to the weights room.
Last, but most of all I want to thank my whole family especially including my parents Hermann and Ilse and of course Oma Anni for their constant support I can always rely on. They made me what I am and I cannot thank them enough. Most of all, I have to thank Annika who was always there for me and always had a shoulder to lean on. She's my little Rock'n Roll! I thank you so much and I like to close this work with the great words of my grandmother

Ein Tropfen Liebe ist mehr wert, wie ein Sack voll Gold!

*(Anni Witte)*